The Textile Institute

# Textile Progress

*Abstracted and Indexed in:*
Compendex
Elsevier Scopus
INSPEC
Textile Technology Index
World Textile Abstracts


*Published on behalf of*
*The Textile Institute by*
*Taylor and Francis*

 Taylor & Francis
Taylor & Francis

Typeset by Aptara, USA

*Textile Progress* is a monograph series that since 1969 has provided critical and comprehensive examination of the origination and application of developments in the international fibre, textile and apparel industry and in its products.

All published research articles in this journal have undergone rigorous peer review, based on initial editor screening and anonymous refereeing by independent expert referees.

Prospective authors are invited to submit an outline of their proposed contribution for consideration by the Editor-in-Chief to: Professor Xiaoming Tao, Editor-in-Chief, Textile Progress, Institute of Textiles and Clothing, Hong Kong Polytechnic University, Hung Hom, Kowloon, Hong Kong. Email: tctaoxm@inet.polyu.edu.hk

# Textile Progress

September 2011
Vol 43 No 3

# Nanotechnology – a new route to high-performance functional textiles

## M. Joshi
## and A. Bhattacharyya

The Textile Institute

Taylor & Francis

Taylor & Francis

Published 2011 by The Textile Institute and Taylor & Francis
2 Park Square, Milton Park, Abingdon, Oxon OX14 4RN
52 Vanderbilt Avenue, New York, NY 10017

ISBN 13: 978-0-415-50603-8 (pbk)
ISBN 13: 978-1-138-45871-0 (hbk)

# CONTENTS

*Textile Progress*
Vol. 43, No. 3, September 2011, 155–233

# Nanotechnology – a new route to high-performance functional textiles

M. Joshi* and A. Bhattacharyya

*Department of Textile Technology, Indian Institute of Technology Delhi, Hauz Khas, New Delhi-110016, India*

(*Received 29 September 2009; final version received 12 June 2010*)

The use of nanomaterials- and nanotechnology-based processes is growing at a tremendous rate in all fields of science and technology. Textile industry is also experiencing the benefits of nanotechnology in its diverse field of applications. Textile-based nanoproducts starting from nanocomposite fibers, nanofibers to intelligent high-performance polymeric nanocoatings are getting their way not only in high performance advanced applications but nanoparticles are also successfully being used in conventional textiles to impart new functionality and improved performance. Greater repeatability, reliability and robustness are the main advantages of nanotechnological advancements in textiles. Nanoparticle application during conventional textile processing techniques, such as finishing, coating and dyeing, enhances the product performance manifold and imparts hitherto unachieved functionality. New coating techniques like sol-gel, layer-by-layer, plasma polymerization etc. can develop multi-functionality, intelligence, excellent durability and weather resistance to fabrics. The present paper focuses on the development and potential applications of nanotechnology in developing multifunctional and smart nanocomposite fibers, nanofibers and other new finished and nanocoated textiles. The four main areas of textile chemical processing, namely nanofinishing, nanocoating, nanocomposite coating and nanodyeing, are covered in the first section of this paper and the second section deals with developments in nanocomposite fibers and nanofibers. The influence of nanomaterials in textile finishing and processing to enhance product performance is discussed. Nanocoating is a relatively new technique in the textile field and is currently under research and development. Polymeric nanocomposite coatings, where nanoparticles are dispersed in polymeric media and used for coating applications, are the most promising route to develop multifunctional and intelligent high-performance textiles. Not much research has been done on applying the concept of nanotechnology in dyeing of textiles except a few reports on dye particle size reduction, structural change in fibers or the surface etching of textiles to create nanostructured surfaces. The reduction in water consumption during nanotechnology applications in textile processing has the potential to control the effluent problems of a textile process house. The most researched area to produce multifunctional, smart fibers is the preparation of nanocomposite fibers where the exceptional properties of nanoparticles have been utilized to enhance and impart several functionalities on conventional textile grade fibers. Nanofibers are gaining popularity in some specialized technical applications such as filter fabric, antibacterial patches and chemical protective suits. Nanotechnological advances in these two areas of nanocomposite fibers and nanofibrous forms have also been reviewed.

**Keywords:** nanofinishing; nanocoating; nanocomposite coating; nanodyeing; nanocomposite fiber; nanofiber

---

*Corresponding author. Email: mangala@textile.iitd.ac.in

ISSN 0040-5167 print/ISSN 1754-2278 online
© 2011 The Textile Institute
DOI: 10.1080/00405167.2011.570027
http://www.informaworld.com

## 1. Introduction

Textiles or textile structures are abundantly used in high-performance technical applications starting from furnishing, food packaging, protective textiles, automotive to aerospace and medical textiles [1]. In order to enhance the performance of these functional textiles, nanotechnology is considered the futuristic approach superceding the conventional chemical, physical or physiochemical modifications. Nanotechnology deals with the science and technology at dimensions of roughly 1 to 100 nanometers (1 nm $= 10^{-9}$ m). Nanotechnology is concerned with materials whose structures exhibit significantly novel and improved physical, chemical and biological properties, phenomena and functionality due to their nanoscale size. 'Nanostructured' materials have grain size less than 100 nm, which leads to a tremendous increase in surface area and high exposure of surface atoms resulting in exceptional physical properties [2]. The mechanical properties of nano-crystalline materials depend on several structural features like grain size and shape, their distribution, pores and their distribution, other flaws/defects and their distribution, surface condition, impurity level, second phase/dopants, stress, duration of its application and temperature [3,4]. The thermal behavior also changes significantly with the change in dimensions at nanoscale [5]. Because of its potential to make changes at fundamental level, nanotechnology is regarded as a key technology that will influence technological development in the near future and also have economic, social and ecological implications [2,6]. Nanotechnology therefore shows significant promise not only in the area of electronics, biomedical, bio-based materials and other advanced materials, human safety and environmental protection but also in consumer and other diversified products, including textiles [2,7].

Although textile industry is a small part of the global research in the emerging areas of nanotechnology, the fibers and textile industries in fact were the first to have successfully implemented these advances and demonstrated the applications of nanotechnology for consumer usage. In the textile field, the research mainly focuses on using nanosize substances and generating nanostructures during manufacturing and finishing processes in order to impart anti-bacterial, water-repellency, soil-resistance, anti-static, flame-retardancy and improved dyeability properties to textiles [8]. The nano-technological approach starts from producing nanofibers or nanocomposite fibers to nanostructure formation and nanoelectronics-embedded garments to produce a wide range of smart and intelligent textiles [9]. The present paper mainly focuses on the possible application of nanotechnology in the chemical processing of textiles covering nanofinishing, nanocoating, nanocomposite coating and nanodyeing, as well in developing nanocomposite fibers and nanofibers with new functionalities that can find application in diverse fields.

## 2. From functional to smart and intelligent textiles

The textiles having some exceptional functionality are defined as high-performance or functional textiles. Functional textiles available in the market are heat-resistant, antibacterial, oil-water repellent, waterproof, windproof with good breathability and moisture transport. They possess optimized material properties like color fastness, tear and rubbing strength, heat and cold resistance etc. Apart from these, there is an emerging class of textiles, namely smart and intelligent textiles. They are able to sense stimuli from the environment, and react and adapt to them by integration of functionalities in the textile structure. Intelligent clothing is generally an article of clothing, footwear or accessories that feature microelectronic sensors. E-textiles, also known as electronic textiles, are fabrics or garments in which computing, digital and electronic components are embedded in the textile structure that allow for the incorporation of built-in technological elements in everyday textiles and clothes. Smart clothes are a combination of intelligent, functional and fashionable textiles. These are generally classified into the following three categories: (1) passive smart textiles, which can only sense the environment; (2) active smart textiles, which can sense the stimuli from the environment and also react to them and (3) smart and intelligent textiles, which not only sense and react but also adapt their behavior to the environment. The responsive behavior or the smartness available in the market are odor release or odor prevention on demand, individually adjustable heat insulation, microcapsules and phase change materials, protection from environmental stress like UV radiation and so forth.

Although encouraging results have been reported in the past in the area of smart textiles, problems such as process complications, mass production, washability and wearing comfort are still under investigation. Studies in the field of intelligent textiles mainly concentrate on developing a wearable-computing device, which integrates smart textures like silicon flexible skins and flexible transistors into regular textiles. Such intelligent textiles can detect environmental conditions, and react and adapt to environmental changes. Smart textiles using fabric-based sensors to monitor gesture, posture or respiration have been exploited in many applications. These smart textiles can measure and monitor the physiological conditions of the wearer. Thus, they could be applied in healthcare systems [10]. During more intense activities as in the case of a sportsperson, the body temperature increases with enhanced heat production resulting in perspiration in order to withdraw energy from the body by evaporative cooling. In order to keep the body temperature within a certain limit, phase change and shape memory materials are used in some intelligent textiles for sportspersons [11]. Phase change materials have high heat of fusion and the transition temperature near to our body temperature and thus absorb or release heat to control the body temperature, whereas shape memory polymers can return from a deformed shape to their original shape induced by external stimuli like temperature, humidity or pH.

The other main application of intelligent materials is to create fantasy design in the fashion field. Some examples of these are music T-Shirts, business garments, solar energy recharge jacket and so forth. Virginia Tech engineering team has designed an E-textile pant, which reduces the risk of falling for older persons (Figure 1). A number of piezo sensors are attached to the knee, ankle and heel of the pant, which constantly monitor the strain on movement of the wearer. Any noise in the signal indicates the irregularity in movement and the system quickly alerts the person to take a corrective action (http://www.physorg.com/news136633890.html).

## 3. Nanotechnology in multifunctional, smart and intelligent textiles

The textile industry is already impacted by nanotechnology. The market for textiles based on nanotechnology is expected to reach US$115 billion by 2012 (http://www.mindbranch.com/Nanotechnologies-Textile-Master-R386-37/) and covers a broad gamut of multifunctional, smart and intelligent textiles. The main research area focuses on using nanosize substances and generating nanostructures during manufacturing and finishing processes [12]. The applications and scope in the area of textiles are huge because of the advancement of nanotechnology in the manufacturing of fibers or yarns, as well as in the development of fabric finishes. Nanosize particles have larger surface area and hence higher efficiency than larger size particles. They do not affect the color and brightness of the substrate. Its main use at present is for optical coatings, where the finer particles give better optical clarity. When a substance is manipulated at sizes of approximately 100 nm, the structure of the processed clothing becomes more compressed. This imparts functionality, which makes clothing stain- and dirt-resistant, saving time and

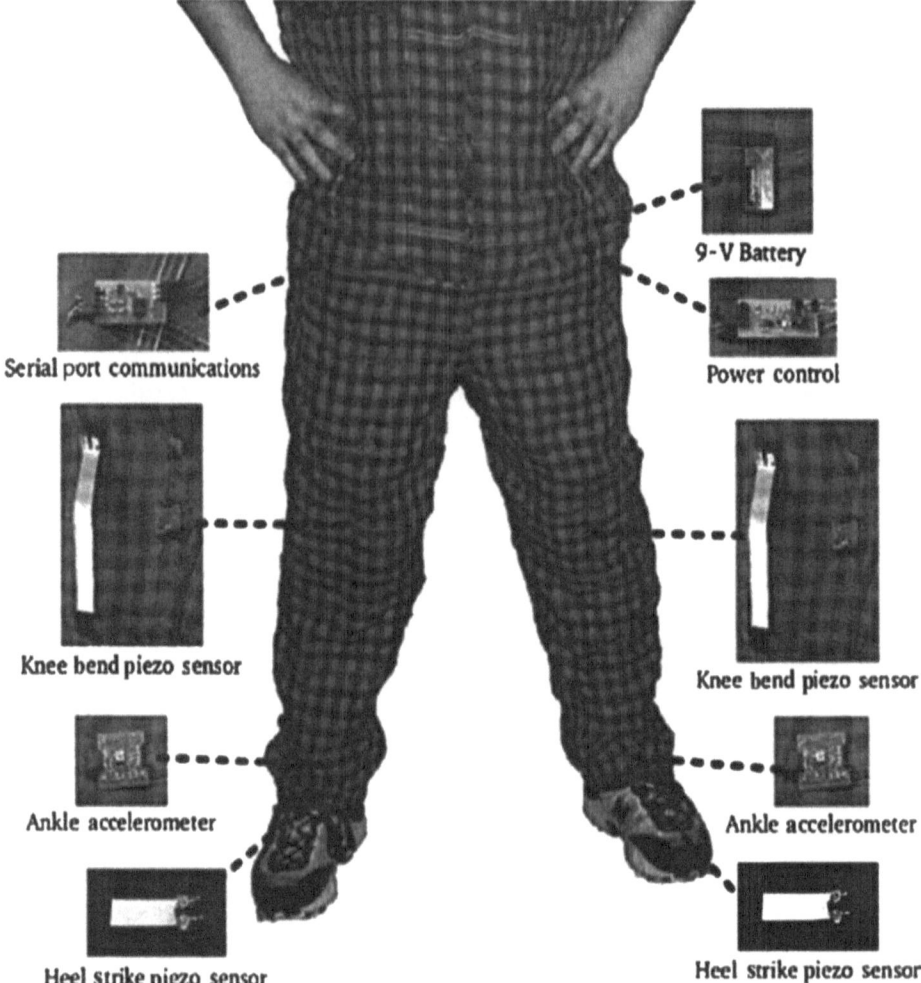

Figure 1. E-textile pants for older persons. Reprinted with permission. *Credit*: Liu et al. © 2008 IEEE. Originally published on PhysOrg.com: http://www.physorg.com/news136633890.html.

laundering cost. Nanofinishes on textiles allow good ventilation and enhanced breathability while maintaining good hand and feel of the original material. Nano-processed garments have protective coatings, which are difficult to detect with the naked eye. Nano-processed products are toxic free, the garments stay bright, fresh looking and are more durable than ordinary processed materials.

Nanotechnology is being used to bring imaginative, exciting and novel properties to textiles for fashion and industry. These range from scent-embedded textiles, stay-clean textiles, textiles with displays and textiles that can change color to bulletproof lightweight textiles. The trend to produce smaller and smaller structures through miniaturization is well known in the manufacturing of microelectronics. The problem of producing wearable computing devices for E-textiles with sufficient comfort can be resolved by using nanotechnology. Such molecular electronic devices can be produced by a technique called atomic force anodization lithography. In this technique a negative voltage applied on a resist solution-coated silicon wafer for inducing field emission from the platinum-coated tip of the atomic force microscopy and resist patterning is created using a constant force on the tip. One such example is functionalized carbon nanotubes attached in different patterns on silicon surface using silane coupling agent, which successfully leads to fabrication of these devices [13]. Such devices hold the key for future integration of electronic technology into the clothing industry. Novel properties of nanoscale materials will make new breakthroughs in a multitude of such technologically important areas.

The developments in the emerging field of nanotechnology-based innovations to create smart and intelligent textiles are being discussed under the broad areas of nanofinishing, nanocoating, nanocomposite coating and nanodyeing in the first subsection. The second subsection covers the developments in the areas of nanocomposite fibers and nanofibers.

### 3.1. Nanofinishing

Textile finishing becomes more thorough, even and precise with innovative techniques of nanotechnology. Nanofinishing or the finishing using nanotechnology can be broadly classified into two major areas – one is the use of nanoparticles in conventional finishing composition, and the other is to use a finishing composition that can develop nanostructures on the surface of the fabric. Both nanoparticles and in situ nanostructures exhibit functionality or multi-functionality with exceptional fastness and without any change in feel and comfort of the fabric. Most of the nanofinishes are in the form of nanoemulsions or nanosols.

*Nanoemulsions*: Nanoemulsions can be defined as emulsions with mean droplet diameters ranging from 50 to 1000 nm. Usually, the average droplet size in nanoemulsions is between 100 and 500 nm, which is smaller compared to conventional emulsion, where the size is in micron range. The terms submicron emulsion and mini-emulsion are used as synonyms. Nanomicelles, nanosols and nanocapsules produced through nanoscale emulsification can more evenly adhere to textile substrate. The study of basic and applied aspects of nano-emulsions is receiving increasing attention in recent years.

The preparation of nanoemulsions requires high-pressure homogenization. Although dispersion or high-energy emulsification methods are traditionally used for nanoemulsion formation, these are also efficiently formed by condensation or low-energy methods. Studies on nanoemulsion formation by low energy methods have shown that the size of the droplets is dictated by the surfactant phase structure (bi-continuous micro-emulsion or lamellar) at the phase inversion point induced by either temperature or composition. The droplet sizes in the range of 20–200 nm and narrow size distributions can be obtained in both the methods.

In contrast to oil in water (O/W) nanoemulsions, which have been studied for many years, the first report on water in oil (W/O) nanoemulsions appeared only five years ago. However, they are currently the object of intensive development. In typical O/W nanoemulsion, the particles exhibit a liquid, lipophilic core separated from the surrounding aqueous phase by a monomolecular layer of surfactant.

The small droplet size, high kinetic stability and optical transparency of nanoemulsions, compared to conventional emulsions, give them advantages for their use in many technological applications. Nanoemulsions, as nonequilibrium systems, present characteristics and properties that depend not only on the composition but also on the method of preparation. Nanoemulsions are obtained by the phase inversion temperature method, phase inversion composition method or self-emulsifying method. Due to the limitations in stability of nanoemulsions, these are prepared just before their use [14]. There are also spontaneous nanoemulsions or thermodynamically stable O/W emulsions, which show superior detergency on textiles [15]. It has been found that the smaller the droplet size, the higher the efficiency.

The use of nanoemulsions as formulations for active delivery and targeting is an active and interesting application. Due to their lipohilic interior, nanoemulsions are more suitable for the transport of lipophilic drugs or enzymes. They support the skin penetration of active ingredients and thus increase their concentration in the skin. Furthermore, nanoemulsions are gaining increasing interest in cosmetics and medicines due to their own bioactive effects. Nanoemulsions are able to favor the transport of suitable lipids into the skin, which increase the barrier function of the skin and thus reduce the epidermal water loss. As an alternative to phospholipid containing nanoemulsions, emulsifier-free O/W submicron emulsions may also be prepared. Using emulsion polymerization, enzymes can be immobilized along with polymers on the surface of cotton fabric. Further, the process can be used to deliver drugs or fragrance in a controlled manner from a fabric [16]. Research is growing in this field to find process optimization, stability and direct application areas [17].

The main application for nanoemulsions is in the preparation of nanoparticles. The majority of publications on nanoemulsion applications deal with the preparation of polymeric nanoparticles using a monomer as the disperse phase (the mini-emulsion polymerization method). In contrast to emulsion and micro-emulsion polymerization, in nanoemulsion polymerization droplet nucleation is reported to be the dominant mechanism, making possible the preservation of size and composition of each droplet. Surfactants containing a polymerizable group are used in nanoemulsion polymerization to protect, stabilize and functionalize the polymer particles. Thus, nanoemulsion droplets can be considered as small nanoreactors.

Commercially available nanofinishes mostly rely on the nanoemulsion techniques to develop nanostructures on the fabric surface. After application of these dispersions by conventional finishing process, the drying and curing of the emulsions lead to durable high-performance nanofinishes on the surface of fabric. The nanoparticle dispersions are also available in the market to impart several functional finishes for the textiles. Some of the commercial nanofinishes launched have been tabulated in Table 1.

The various nanofinishes being researched to impart a range of functionalities based on the property achieved in the final textile products are described in this subsection.

### 3.1.1. Oil and water repellent nanofinishes

In order to generate oil and water repellent surfaces in cotton, low surface energy materials are used. A number of water and oil repellent finishing chemicals are available commercially that could be classified into fluorocarbon containing or non-fluorocarbon-based finishes [19]. Silicon-based waterproofing is also used either as such or in combination with fluorocarbon-based agents. Fluorinated compounds are the most popular ones. However, the current market demands are for non-fluorocarbon-based finishes because the fluorinated materials are expensive in price and are often under attack due to growing environmental consciousness and increasing strict legal regulations. Besides this, the durability of finish and retaining the original feel and strength of the fabric are also important criteria. Nanotechnology-based novel and innovative nanofinishes seem to be closer to achieving these consumer demands.

The premier range of Nano Care® and NanoPel® nanofinishes marketed by the US Company, NanoTex Inc., are the next generation easy care finishes based on nanotechnology. These finishes, which come under the Resist Spills™ Category protect the fabric against both water and oil-based liquid stains/soils (www.nano-tex.com). Nanocare fabrics are created by modifying the outer surfaces of the cylindrical cotton fibers with a fuzz of minute whiskers, which creates a cushion of air around the fiber. The whiskers create fewer

Table 1. Commercially available functional nanofinishes [18]. Reprinted from M. Joshi, *The impact of nanotechnology in polyesters and polyamides*, in *Polyesters and Polyamides*, B.L. Deopura, R. Alagirusamy, M. Joshi and B. Gupta, eds., Woodhead, Cambridge, UK, 2008, pp. 354–415, with permission from Woodhead Publishing Ltd., Cambridge, UK (www.woodheadpublishing.com).

| S. No. | Name of the finish | Company | Properties | References |
|---|---|---|---|---|
| 1. | Resists Spills™ | Nano-Tex | Water-repellent and stain resistant. | www.nanotex.com |
| 2. | Coolest comfort™ | Nano-Tex | Impart superior wicking properties to a previously hydrophobic synthetic or resin-treated cotton. | www.nanotex.com |
| 3. | NanoCare® | Nano-Tex | Helps stains to wash out easily. | www.nanotex.com |
| 4. | Zonyl® | DuPont | Water-repellent fluorinated polyurethanes as soil-release finish. | www.dupont.com |
| 5. | Zyvere® | Nanovere Technologies | Self-cleaning paint surfaces. | www.nanovere.com |
| 6. | NanoSphere® | Schoeller Textil AG | Repels water and oil drops and prevent soil particles from attaching themselves on the fabric surface. | www.schoeller-textiles.com |
| 7. | Ultra-Fresh™ Silpure | Thomson-Research Associates | Antimicrobial treatment for textiles. | www.ultrafresh.com |
| 8. | NanoArc® Zinc oxide | Nanophase Technologies Corp. | Excellent UV protection, it does not degrade over time. | www.nanophase.com |
| 9. | NewPro Nano Textile | Newpro | Anti-adhesive coating, strong hydro-oleo phobic, stain resistant, high efficiency. | www.g-pro.com |
| 10. | NANO | NANOBIZ.PL Ltd. | Waterproof nanocoating for home textiles and leathers. | www.nanoprotect.co.uk |
| 11. | Z-Cote® | BASF Corp. | Clear solution for skin protection against UV light. | www2.basf.us |
| 12. | Nano anion powder | Shanghai Huzheng Nano Technology Co. Ltd. | Metal oxide powder for antistatic finish preparation or can be added during fiber spinning. | www.hznano.com |
| 13. | Anti pollen fabric | Miyuki keori Co. | Nano-infused materials, 30-nm diameter antistatic polymer nanoparticles attach to fiber and prevents pollen from attaching. | http://gizmodo.com/101976/anti+pollen+fabric |
| 14. | NYACOL® A1540N | Nyacol Nano Technologies Inc. | Colloidal antimony pentoxide fire retardant finishing. | www.nyacol.com |

points of contact for dirt. When water is applied to soiled fabric, the dirt adheres to the water far better than it adheres to the textile surface and is carried off with the water as it beads up and rolls off the surface of the fabric. As the attached whiskers are of nanoscale size, they do not affect the hand and breathability of fabric and can withstand 50 home launderings [20,21]. The Resist Spills™ protection is applied to fabrics during finishing stage simply by padding or dip or dry coating. This nanofinish can be applied to natural fibers such as cotton, wool and silk, as well as synthetics such as polyamides, polyesters, acrylic and so forth. Leading garment manufacturing brands such as Burlington, Galey & Lord, Dan River, Eddie Bauer and Lee have used this licensed technology for their Nanocare® range of commercial products. The large-scale production of shirts and shirting fabrics has also been licensed to several South Asian countries, including India. Indian companies like Arvind Mills and Madura Coats have already licensed some of these technologies to produce nanofinished garments.

Another interesting observation reported is the use of organo-modified nano-clays influencing the curing behavior of the fluoroelastomers used for giving heat, oil and solvent resistant finish. They show a higher rate and state of cure when nanoclays are incorporated into them [22]. Thus, the curing conditions of the oil and water repellent finish can be modified and more oil and water repellent effect with less amount of fluorocarbon application can be achieved through the use of nanoclays.

### 3.1.2. Super hydrophobic nanofinishes

In 1990, Wilhelm Barthlott discovered that the lotus plant owes the property of self-cleaning to the high density of minute surface protrusions (Figure 2). This is known as

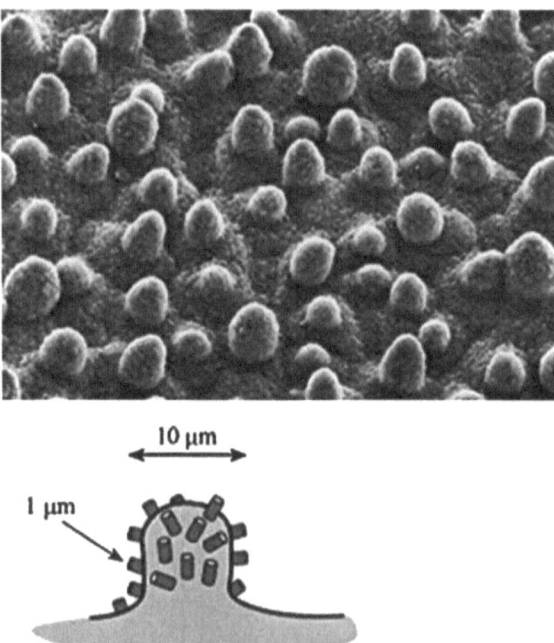

Figure 2. Close up of a lotus leaf. Reprinted with permission from http://nanotechweb.org/cws/article/tech/21936 (related to Y.-T. Cheng and D. Rodak, Appl. Phys. Lett. 86 (2005) p. 144101). Copyright 2005, American Institute of Physics.

'Lotus Effect' that possesses super-hydrophobic surfaces and self-cleaning phenomenon due to low surface energy materials and nanoscale roughness in the leaf structure. These protrusions catch deposits of soil preventing them from sticking. When it rains, the water rolls around as droplets, removing dust as it moves.

This self-cleaning property can be developed as a technological innovation when reproduced on the surface of woven fabrics. Thus, produced super-hydrophobic surfaces using nano technological process have drawn great attention for both fundamental research as well as practical applications [23]. These fabrics will have specific applications such as sails, shelter fabrics or certain garments. It is well understood that the hydrophilic and hydrophobic surfaces are governed by both surface roughness and chemical composition. Conventionally, modification of surface chemistry is always in conjunction with enhancement of the surface roughness in order to form the super-hydrophobic films. Super-hydrophobic surfaces can be prepared by controlling the surface topography by various processing methods, such as sol–gel method, organic/inorganic hybrid method, CVD method, electrochemical method, embossing method, plasma method, phase separation method, template and other methods [24].

Super-hydrophobic silica-based surfaces have been prepared by the sol–gel process based on silica nanoparticles and perfluoro-octylated quaternary ammonium silane coupling agent (PFSC) using cotton fabrics as a substrate. PFSC was synthesized and appropriate particle size of silica sol was prepared successfully. The silica sol and PFSC were applied to the cotton fabrics by conventional pad-dry-cure process. The fabrics treated with both silica sol and PFSC showed high hydrophobicity and oleophobicity due to the surface modification by the silica sol. This is a combined technique to introduce nanoroughening and lower down its surface energy with the help of two components mimicking the lotus leaf. Silica nanoparticles make the textile surface much rougher, and perfluoro-octylated quaternary ammonium silane coupling agent lowers the surface free energy [25]. Textiles coated with this coating showed excellent water repellent property and water contact angle (CA) increased from 133° on cotton fabrics treated with pure PFSC without silica sol pretreatment up to 145°. The oil repellency was also improved to 131° from 125° with pure PFSC.

Another process for the preparation of superhydrophobic silica-based surfaces is by adding polypropylene glycol (PPG) polymer into the silica precursor and subsequently, by removing the PPG at 500°C, surface roughness is obtained (Figure 3).

In this case, hexamethyldisilazane (HMDS) is used to bond hydrophobic groups onto the films. Physical properties of the as-prepared films were analyzed by contact angle measurements; scanning electron microscopy (SEM), UV–VIS scanning spectrophotometer and Fourier transform infrared (FT-IR) spectrophotometer. The experimental parameters were varied by the type of silane species, the weight ratio of the PPG solution to precursor solution, the hydrolysis time of the precursor solution, the molecular weight of PPG, the casting temperature and the evaporation temperature. It was found that the contact angles of the films could be controlled by the type of silane species, the hydrolysis time of the precursor solution, the molecular weight of PPG, the casting temperature and the evaporation temperature. Nevertheless, preparation of the films at lower evaporation temperature resulted in the high contact angles of the as-prepared films because the surface roughness enhanced at very low temperature due to the increased microphase separation of the PPG polymer. The films prepared at 5°C evaporation temperature exhibited excellent contact angles, which were greater than 160° when the weight ratio of the PPG solution to the precursor solution was 5. The optical transmittance of such films was also excellent (greater than 80%) [24].

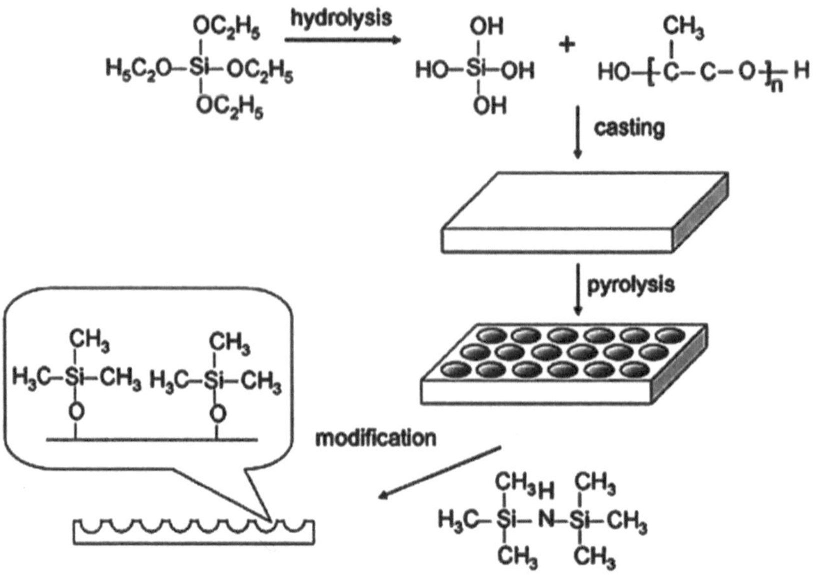

Figure 3. Schematic of the procedure on the preparation of superhydrophobic films [24]. Reprinted from K.C. Chang, Y.K. Chen and H. Chen, Surf. Coat. Tech. 201 (2007) pp. 9579–9586, with permission from Elsevier.

### 3.1.3. Photocatalytic self-cleaning nanofinishes

Daoud and Xin [26–28] of the Hong Kong Polytechnic University's Nanotechnology Centre for Functional and Intelligent Textiles and Apparel have developed a process for the sol-gel coating of nano-$TiO_2$/nano-ZnO particles on textile substrates at low temperature. They claimed that photocatalytic self-cleaning properties could be imparted to the coated cotton fabric on coating with titanium dioxide ($TiO_2$) nanoparticles (about 20 nm in size). Transparent thin film coatings of sol–gel-derived $TiO_2$ are produced on cotton fabrics at a low temperature of 150°C. X-ray diffraction (XRD) patterns revealed the existence of anatase phase in small scale within the titania layers. Anatase nanocrystalline titania coatings can be developed on cotton fabrics from alkoxide solutions using a low-temperature sol–gel process under ambient pressure. Titania coatings of the anatase form were obtained through a classical hydrolysis and condensation reaction of titanium isopropoxide that was followed by a boiling water treatment for 180 min.

Spectroscopic and microscopic characterizations of the titania thin films showed that the anatase form is predominant throughout the film after the boiling water treatment and the size of the grains is about 20 nm. SEM images also show the formation of continuous layers of $TiO_2$ with grains of about 20 nm in diameter. The photocatalytic activity of the formed layers was studied by means of their antibacterial property. The UV-blocking effect of titania coatings was also studied. The titania coating on fabrics is a promising process for use as an antibacterial photocatalyst in the textile industry.

The coated fabrics show discoloration of red wine and coffee stains on irradiation of samples in a sun test solar simulator. The photocatalytic activity of semiconductor oxides, such as $TiO_2$, is attributed to promotion of an electron from the valance band to the conduction band brought about by the absorption of a photon of ultra band gap ($\sim$3.2 eV) light whose energy (h$\upsilon$) is greater than the energy difference between

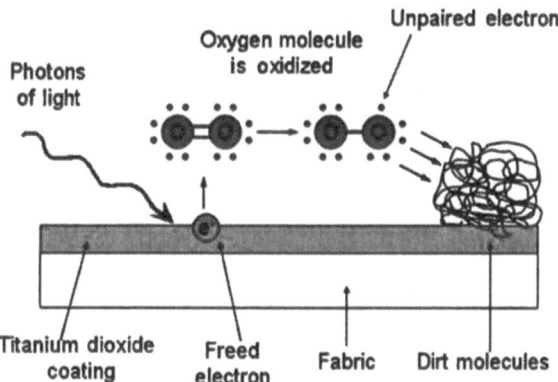

Figure 4. Self-cleaning mechanism of cotton fabric with TiO$_2$. Reprinted from http://nanopedia. case.edu/NWPage.php?page=how.self.clean.fabric.works, with permission from Dave Smith (ed.); accessed 15 February 2011.

electrons in the valance band and the conduction band. When photons of light hit the TiO$_2$ layer, electrons are excited up to the conduction band, which react with oxygen molecules in the air (Figure 4). The double bond of the oxygen molecule is broken (http://nanopedia.case.edu/NWPage.php?page=how.self.clean.fabric.works) and in presence of oxygen and/or water, super-oxide (O$_2$) and/or hydroxyl (OH) radicals are formed, which attack adsorbed organic species on the TiO$_2$ surfaces and decompose them [29].

TiO$_2$ can be applied to cotton fabric by padding or coating method. But the coated fabric is less suitable than TiO$_2$-padded cotton fabric for self-cleaning and ammonia pollution control activity [30]. This is because of the binder used in coating that reduces the exposure of TiO$_2$ nanoparticles on the fabric surface and thus reduce its activity. The 1:1 mixture of TiO$_2$ and SiO$_2$ colloids leads to an organized structure of highly dispersed TiO$_2$ particles always surrounded by amorphous silica during the dip-coating and subsequent thermal treatment on cotton [31]. The TiO$_2$–SiO$_2$ layer thickness on the cotton fibers is in the range of 20–30 nm. The photocatalytic activity of such TiO$_2$–SiO$_2$-coated cotton textiles to discolor red wine is more efficient than TiO$_2$-coated ones.

### 3.1.4. Hydrophilic nanofinishes

The poor moisture absorption property of synthetic fabrics like polyester and polyamides limits its applications in the apparel sector. The new range of hydrophilic nanofinishes 'Cotton touch'$^{TM}$ and 'Coolest Comfort'$^{TM}$, commercialized by NanoTex, USA, makes the synthetic fabric look and feel like cotton (www.nano-tex.com). These finishes last 50 launderings.

'Cotton touch'$^{TM}$ is a naturally soft fabric enhancement designed to make synthetic fabric look and feel like cotton. It is available as 'Nano-touch'$^{®}$ and gives a durable cellulose wrapping over synthetic fibers (Figure 5). Cellulosic sheath and synthetic core together form a concentric structure to bring overall solutions to the drawbacks of synthetics such as static discharge, harsh handle and glaring luster [32]. 'Coolest comfort'$^{TM}$ provides breakthrough moisture wicking to draw moisture away from the body while drying quickly. The popular product in this category is 'Nano-dry'$^{®}$ (Figure 5). It improves the moisture absorption of polyamides and polyesters, making them hydrophilic and comfortable. The

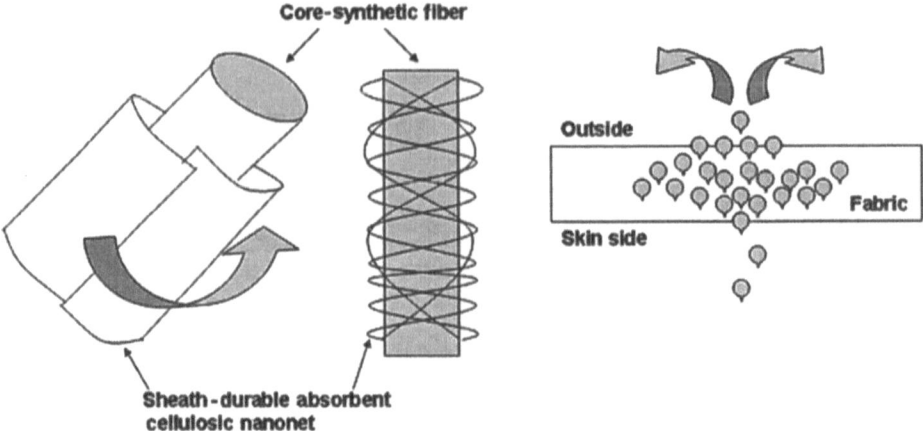

Figure 5.  The schematic diagram of the concept of 'Coolest comfort'[TM].

main applications are in sportswear and close to body garments that require perspiration absorbency.

### 3.1.5. Antibacterial nanofinishes

Most of the textile fabrics are not resistant to pathogenic bacteria or fungi whether the constituent fiber is natural or synthetic. The humidity, sweat and moisture create a suitable environment for the growth of a range of microbes (pathogenic or non-pathogenic) on the textile inner or outer wear. These can cause diseases, bad odor for the user and loss of fabric properties such as color or strength. So various antibacterial finishes and disinfection techniques have been developed for all types of textiles. Nowadays, the control of microorganisms on textile fabrics is not only limited to the medical textiles but also extended to everyday household. For a long time, the chemical agents, such as halogen ions, were used to control microorganisms. Many of these agents have been employed for generations, while several new antibacterial agents for textiles based on metal salt solutions have been developed recently. But pure metals have not been used normally for antibacterial finishing. Heavy metals are usually toxic and very reactive with proteins. They are believed to bind protein molecules, which inhibit the cellular metabolism thus leading to death of microorganisms. The high level of bacterial resistance to obtain the desired quality products by simple application techniques at low processing costs is the commercial interest of many textile finishers. It is here that the nanotechnology comes into play as new antibacterial finishing has been developed using simple methods by taking advantages of extraordinary physical and chemical advantages of nanomaterials. The multidisciplinary approach of investigations and researches on nanotechnology leads to integrate several bioactive components on textile fabrics in order to impart antibacterial, antimicrobial or antifungal property. The main interest for a material scientist is the fact that nanostructured materials have a higher surface area than conventional materials. For this reason, a small amount of noble metal nanoparticles can be applied to the surface of fibers by padding of colloidal solution and inhibit the growth of microorganisms.

The most prominent nanoproduct as an antiseptic and disinfectant being used for medical applications is nanosilver. Though the mechanism is not yet fully understood, silver's disinfectant property for hygienic and medicinal purposes is time honored and prominent along with its many applications. Silver is known for its antibacterial properties against bacteria and microorganisms. Silver vessels were used in ancient times to preserve water and wine and silver powder was believed to have beneficial healing and anti-disease properties and listed as a treatment for ulcers. Mainly the silver compounds were actually employed in medical practice. Silver compounds were major weapons against wound infection in World War I until the advent of antibiotics. In 1884 German obstetrician C.S.F. Crede introduced 1% silver nitrate as an eye solution for prevention of Gonococcal ophthalmia neonatorum, which is perhaps the first scientifically documented medical use of silver. Further, topically used silver sulfadiazine cream was standard antibacterial treatment for serious burn wounds and is still widely used in burn units. Due to the problem of argyrosis (a disease causing irreversible pigmentation of skin and eyes) after prolonged exposure to silver or silver compounds and with the advent of antibiotics, silver's luster largely faded away as an anti-infection agent. However, advancement of modern science especially nanoscience has helped silver renew its lost luster. Metallic silver is subjected to new nano engineering technologies with resultant extraordinarily novel morphologies and characteristics i.e. metallic silver is now engineered into ultrafine particles whose size is measured in nanometres (nm). Nanosilver particles are generally smaller than 100 nm and contain 20–15,000 silver atoms, depending on their size. At nanoscale, silver exhibits remarkably unusual physical, chemical and biological properties. Great research efforts have been committed to this respect and yielded exciting and encouraging results. As a consequence, applications of silver nanoparticles or nanosilver have been widely explored in the healthcare sector.

The nanoscale silver particles are produced or dispersed in colloidal solution before applying on the textile fabrics and their antibacterial effect and durability is found to be excellent. Silver compound or nano particle can be used as a durable antimicrobial finish if it is encapsulated with a fiber reactive polymer like polystyrene co-maleic anhydride. Due to its strong antibacterial activity, nanosilver is successfully applied on various natural and synthetic fabrics. Silver nanoparticles incorporated polypropylene (PP) exhibits superior antibacterial activity relative to the PP containing micron-sized particles [33]. The silver has been incorporated into the PP by melt mixing in twin screw extruder and fibers are prepared. The fibers are found to be reasonable in terms of mechanical properties and it has been found that more than 5 times loading of micron sized silver particles are required to have the same antibacterial effect produced by nano silver against both gram positive and gram negative bacteria. The colloidal solution of nanosilver shows strong antibacterial activity in cotton and polyester fabrics even after laundering several times [34]. Wool fabric exhibits moth-proofing, antibiotic and antistatic properties when padded with sulfur nanosilver ethanol based colloids having a very low silver content (20 ppm) [35]. Scanning and transmission electron microscopy (SEM and TEM) studies on the biocidal action of silver nanoparticles show an effective bactericide activity of the silver nano particle. The results confirmed that the *E. coli* cells were damaged with interaction to the nanosilver, showing formation of "pits" in the cell wall of the bacteria, while the silver nanoparticles were accumulated in the bacterial membrane (Figure 6b). A membrane with such morphology exhibits a significant increase in permeability, resulting in death of the cell. The broad spectrum anti-microbial activities of nanosilver are now acknowledged in colloidal solution which can be applied as a nanofinishing on fabric or fiber surfaces.

According to a market research report, "Silver nanoparticles are emerging as one of the fastest growing product categories in the nanotechnology industry". The remarkably strong anti-microbial activity is the major direction for development of nanosilver products. A wide category of products based on silver nanoparticles is already available in the market.

Figure 6. (a) Silver coated antimicrobial fibers (SEM image) and (b) nanosilver attack on bacteria cell wall causing damage on the wall [36]. Reprinted from I. Sondi and B. S. Sondi, J. Colloid Interf. Sci. 275(2004) pp. 177–182, with permission from Elsevier.

In the medical arena, there are wound dressings, contraceptive devices, surgical instruments and bone prostheses all coated or embedded with nanosilver. In daily life, consumers may have nanosilver containing room sprays, laundry detergents, water purificants and wall paint. Silver nanoparticles are also incorporated into textiles for manufacture of clothing, underwear and socks. Washing machines also employ nanosilver. More nanosilver products are in the pipeline. It is estimated that among all the nanomaterials in the medical and healthcare sector, nanosilver application has the highest degree of commercialization. Some commercially available nanosilver based antibacterial products for textile finishing available in the market are shown in Table 2.

As exposure to nanosilver in the human body is becoming increasingly widespread and intimate, consequently, silver in the form of nanoparticles has gained an increasing access to tissues, cells and biological molecules within the human body. The traditional belief is that except for argyrosis and some minor problems, silver is relatively non-toxic to mammalian cells. Silver poisoning only occurs among workers who have chronic history of silver exposure. Metallic silver was viewed to be a minimal health risk. However, once reaching nanoscale, certain materials do exhibit significant toxicity to mammalian cells even if they are biochemically inert and biocompatible in bulk size. The carbon nanotube is the finest example of such cases. As the use of nanosilver is becoming more and more widespread in medicine and related applications, toxicological and environmental issues related to the use of silver in nano form are being raised. Bacterial cells constantly exposed to stressful situations develop an ability to resist these stresses, essential for their survival. The ability of microorganisms to grow in the presence of metal might result from specific mechanisms of resistance. Such mechanisms include alteration of chemical structure and or toxicity by changes in the redox state of the metal ions. However, silver has been well known as non-toxic in spite of claiming to kill many different disease organisms. In literatures, silver is skin

Table 2. Commercially available nanosilver-based antimicrobial products [18]. Reprinted from M. Joshi, *The impact of nanotechnology in polyesters and polyamides*, in *Polyesters and Polyamides*, B.L. Deopura, R. Alagirusamy, M. Joshi and B. Gupta, eds., Woodhead, Cambridge, UK, 2008, pp. 354–415, with permission from Woodhead Publishing Ltd., Cambridge, UK (www.woodheadpublishing.com).

| Sl. No. | Company | Product name | Type of product | References |
|---|---|---|---|---|
| 1. | NanoHorizons | SmartSilverTM | Additive. | www.nanotech-now.com |
| 2. | ABCNanotech | SARPU | Finishing/coating solution (Nanosized silver dispersed in liquor). | http://www.abcnanotech.com/ |
| 3. | AcryMed | SilvaGard® | Solution surface treatment. | http://www.acrymed.com |
| 4. | Advanced Nano Product (ANP) | Silver nanopaste | Powder/coating solution. | http://www.anapro.com/korean/default.asp |
| 5. | JR Nanotech | SoleFreshT | Dispersion containing 0.3% w/w nano-silver. | http://www.jrnanotech.com |
| 6. | Nanocid | Silver nanoparticles | Powder/colloid. | http://www.nanocid.com/index.html |
| 7. | NanoGap | Silver nanoparticles | Powders or inorganic or aqueous solutions. | http://www.nanogap.es |
| 8. | Arc Outdoors | 'X-Fiber' | Fibers and fabrics with nanosilver-based antimicrobial properties. | http://www.arcoutdoors.com/technology.html |
| 9. | Shanghai Huzheng Nano Technology Co., Ltd. | Nano-silver spray | Antimicrobial self-cleaning can be applied on ceramic, plate, glass and fiber. | www.tradeindia/selloffer/1201107/antimicrobial-anion and infrared-multifunctional-paint |
| 10. | NANOBIZ.PL Ltd. | NANOVIS | Nano silver-based solution, removes odor and bacteria. | www.nanoprotect.co.uk |
| 11. | Silverlon Inc. | Silverlon® | Nanosilver-based wound care dressings. | www.silverlon.com |
| 12. | Smith & Nephew | Acticoat | Nanosilver-based wound care dressings. | www.acticoat.com |
| 13. | Miji Tech. Co. | Ag Nano powder | Antibiotic powder. | www. mijitech.en.ec21.com |
| 14. | AgActive | AgActive antibacterial spray | Prevents the development of antimicrobial resistance commonly encountered with many traditional disinfectants and antibiotics. | www.agactive.com |

friendly and does not cause skin irritation but nanosilver is not so innocent. In sharp contrast to the attention paid to new applications of nanosilver, only a few studies are available on the interaction of nanosilver particles with the human body after entering via different portals. Bio-distribution, organ accumulation, degradation, possible adverse effects and toxicity of the nanosilver are recognized by various groups of scientists [37]. In the human body, silver nanoparticles can undergo a series of processes like binding and reacting with proteins, phagocytosis, deposition, clearance and translocation. On the other hand they can initiate a spectrum of tissue responses such as cell activation, generation of reactive oxygen species, inflammation and cell death. It has also been reported that the silver nanoparticles leads to the release of toxic nanosilver to freshwater ecosystems which may harm the aquatic life [38]. The toxicity related to nanosilver applications needs to be addressed carefully as nanosilver based products get into consumer products including textiles.

### 3.1.6. UV protective nanofinishes

The most important function performed by the UV protective garment is to protect the wearer from the weather and the harmful ultraviolet radiations of the sun. The effect of weathering on the textiles used in agriculture and horticulture is of paramount importance as they are exposed to natural weathering for longer periods. The UV-blocking property of a fabric is enhanced when a dye, pigment, delustrant, or ultraviolet absorber finish is present that absorbs ultraviolet radiation and blocks its transmission through a fabric [39]. Fabric treated with UV absorbers ensures that the clothes deflect the harmful ultraviolet rays of the sun, reducing a persons UV exposure and protecting the skin from potential damage. The extent of skin protection required by different types of human skin depends on UV radiation intensity and distribution in reference to geographical location, time of day, and season. This protection is expressed as Sun Protection Factor (SPF), the higher the SPF value the better the protection against UV radiation. Metal oxides like zinc oxide (ZnO) as UV-blocker are more stable when compared to organic UV-blocking agents. Nano ZnO enhances the UV-blocking property due to its increased surface area and intense absorption in the UV region. Besides, ZnO nanoparticles scores over nano-silver in cost-effectiveness, whiteness, and UV-blocking property [12]. ZnO nanoparticles can be prepared by the wet chemical method using zinc nitrate and sodium hydroxide as precursors and soluble starch as the stabilizing agent. These nanoparticles, having an average size of 40 nm, were padded on the bleached cotton fabrics using acrylic binder. About 75% UV blocking was recorded for the cotton fabrics treated with very low (2%) concentration of ZnO nanoparticles. In the case of nano-ZnO coated fabric, due to its nano-size and uniform distribution, friction was significantly lower than the bulk-ZnO coated fabric, hence the nano-ZnO coated fabric also possesses a superior feel [40]. Further studies are under way to evaluate wash fastness, antimicrobial properties, abrasion properties and fabric handle properties. BASF has commercialized nano-ZnO dispersion for UV protection as Z-Cote® finish which is completely transparent and has a long lasting sun blocking capacity (Table 1).

### 3.1.7. Antistatic nanofinishes

Synthetic textiles are prone to static charge accumulation as they absorb less water. It has been reported that nanosized $TiO_2$, ZnO whiskers, nanoantimony-doped tin oxide and silane nanosol could impart antistatic properties to synthetic fibers [41–43]. $TiO_2$, ZnO and $TiO_2$ nanoparticles are electrically conductive materials and help to dissipate the static charge in these fibers. These kind of conducting nanofillers are successfully used in different finishes to reduce the static charge build on synthetic fabrics (Table 1). The functional groups responsible for hydrogen bond formation reduce the static charge accumulation by increasing the moisture content of the fabrics. Silane nanosol finish improves antistatic properties as the silane gel nanoparticles absorb moisture due to $-NH_2$ and $-OH$ groups present in them. W. L. Gore and Associates have used nanotechnology to develop an antistatic membrane for protective clothing. Goretex®antistatic is a multifunctional textile that protects the wearer from electrostatic discharges, weather, heat and flame (http://www.emeraldinsight.com/journals.htm?articleid=875516&show=html). Electrically conductive nanoparticles are homogeneously anchored in the fibers of the Goretex membrane, creating a durable electrically conducting network that prevents the accumulation of static charge.

### 3.2. Nanocoating

Materials are coated for a number of reasons: coatings can make a substance biocompatible, increase a material's thermal, mechanical, or chemical stability, increase wear protection, durability, or lifetime, decrease friction or inhibit corrosion, or change the overall physico-chemical and biological properties of the material. Coating on a wood, metal, textile, leather substrate with a polymeric layer is an age old practice, carried out to impart specific surface properties like shine, wear, hydrophobicity, water and gas barrier, conducting, antistatic, antibacterial etc. But conventional coating has several problems like strength loss, improper adhesion, poor abrasion resistance and less durability [44,45]. To achieve a desired amount of surface property, higher coat to weight ratio is often required. Researchers focus their interest to overcome these problems by minimizing coat to weight ratio. With the advent of nanotechnology, a new area has developed in the realm of textile nanocoating, which is really very thin (~50 nm). Coating is simply the act of covering a material with a layer; hence, nanocoating is either to cover with a layer with thickness on nanometer scale or to cover a surface with a nanoscale entity. New approaches include creating nano-structured surfaces with significantly optimized or enhanced properties. Nanostructures develop by ordering of the components into the desired coating formation or develop during the coating process. Although nanocoatings on fibers or textiles are to date a relatively unexplored area with very few cited reports [46–48], these nanostructures impart properties that have significant impact on coating reactivity, corrosion resistance, strength and durability [49].

Nanoscience and nanotechnology also have the potential to break some of the old paradigms about coating processes and the desired structures of coatings to create new types of highly functional systems. Nanocoating techniques lead to the synthesis of novel functional materials with properties that depend on the combination of components employed in the fabrication process. Catalysis, electronics, biomaterials engineering, and materials chemistry are areas that will benefit from designed materials with specific functionality, and controlled arrangement or organization of the pore structures and particles, and their size. The rapid growth of coating and dispersion technology using nanoparticles to improve the properties of the substrate has seen tremendous advances over the past decade. These advances cover the spectrum from scientific achievements resulting from long-term research to commercial successes. It will not be an exaggerated statement that the future of textile coatings is nanocoating as this is the ideal method, at least theoretically, in respect of coat weight; improving or imparting functionality with smallest change in base properties of substrates. There are numerous widely used nanocoating procedures including vapor deposition, plasma-assisted/ion-beam-assisted techniques, chemical reduction, pulsed laser deposition, mechanical milling, magnetron sputtering, self-assembly, layer-by-layer coating, dip

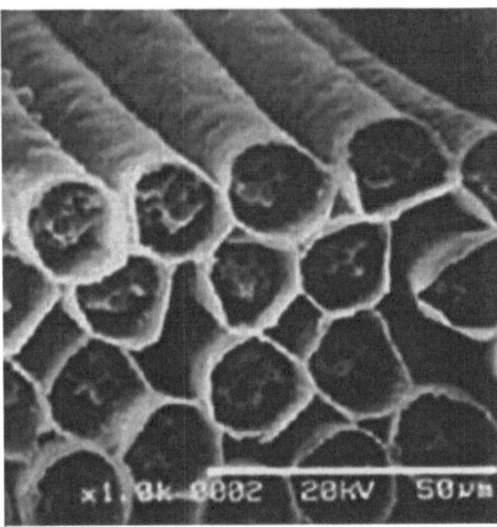

Figure 7. Nanomatrix Technology from Toray for nanocoatings on textiles through self-assembly (http://www.voyle.net/Nano%20Textiles/Textiles-2004-003.htm). Reprinted from http://www.voyle. net/Nano%20Textiles/Textiles-2004-003.htm with permission from David Voyle.

coating, sol-gel coating, and electrochemical deposition. The majority of these techniques are preferentially used to coat planar substrates. The most important techniques considering suitability of application on the uneven surface of a textile substrate such as self-assembly, sol–gel, plasma polymerization, layer-by-layer and some others are described below.

### 3.2.1. Self-assembly based nanocoating

Toray Industries, Inc. have succeeded in developing a 'nano-scale processing technology' that allows the formation of molecular arrangement and molecular assembly necessary to bring out further advanced functionalities in textile processing (http://www.voyle.net/Nano%20Textiles/Textiles-2004-003.htm). This 'nano-scale processing technology' named 'NanoMATRIX' forms the functional material coating (10–30 nm) consisting of nano-scale molecular assembly on each of the monofilaments that forms the fabric (woven/knitted fabric) (Figure 7).

'Nano-matrix' is based on the concept of 'self-organization' by controlling the conditions like temperature, pressure, magnetic field, electrical field, humidity, additives, etc. It is possible to control the state of molecular arrangement and/or assembly of functional materials on each of the monofilaments in nanoscale sizes precisely by controlling the interaction and responses between the functional material to be coated and the fiber material (polymer). The application of this technology is expected to lead to the development of new functionalities as well as remarkable improvements in the existing functions (quality, durability, feel, etc.) without losing the fabric's texture.

### 3.2.2. Plasma polymerization assisted nanocoating

Plasma is a partially ionized gas, composed of highly excited atomic, molecular, ionic and radical species with free electrons and photons. Sufficient additional energy, supplied to gases by an electric field, creates plasma. The reactive species of plasma, resulting from ionization, fragmentation, and excitation processes, are high enough to dissociate a wide variety of chemical bonds, resulting in a significant number of simultaneous recombination mechanisms. Plasma surface treatments show advantages of modifying the surface properties of inert materials. The main advantages of plasma polymerization methods are: (1) applicability to almost all organic, organo-metallic and hetero-atomic organic compounds, (2) modification of surface properties without altering the bulk characteristics, (3) low quantities needed of monomeric compounds making it non energy intensive, and (4) wide applicability to most organic and inorganic structures [50]. Plasma reduces air, water and land pollution in comparison to conventional methods of wet chemistry. However, during the plasma treatment process gases like fluorine, ammonia and nitrous oxide are used to develop certain properties and the waste gases liberated from the process are hazardous. Plasma can also lead to stiffening and degradation of fabrics. Vacuum glow discharge plasma is not economic for textiles and atmospheric pressure plasma equipments are not so consistent over a width of fabric (approx. 1.5–2.0 m). For the treatment of fabrics, cold plasma (near room temperature) is used. But cold plasma treatments require the development of reliable and larger systems to upscale for commercial applications.

In textiles, plasma is used mainly for four basic processes: *cleaning* by removing organic contaminates from surface of fabric, *activation* of the fabric surface by anchoring functional groups, *grafting* of monomers like acrylic acid on fabric surface and *deposition* of thin polymeric films. Characteristics that can be improved through plasma processes include wettability, flame resistance, adhesive bonding, printability, electromagnetic radiation reflection, surface hardness, hydrophilic-hydrophobic tendency, dirt-repellent and antistatic properties. After plasma treatment the surface energy of polymeric surface increases and wettability and adhesion enhancements are produced even for fluoropolymers. Depending on the gas employed, plasma treatment may render the surface very hydrophilic or hydrophobic. Generally, inert gases or oxygen plasmas are used for the plasma-cleaning process. Activation plasma processes provide functional groups when treated with oxygen, ammonia, nitrous oxide or sulfur dioxide to almost any polymeric surface in order to significantly enhance the adhesion characteristics. In grafting, an inert gas such as argon is employed as the process gas to create free radicals on the textile surface and then a monomer is introduced into the chamber to become grafted. This is a procedure for low-pressure plasma treatment but grafting can be obtained with atmospheric plasma processing too. When a more complex molecule is employed as the process gas, plasma-enhanced chemical-vapor deposition occurs onto the polymer surface [51]. Plasma polymerization or the deposition of solid polymeric materials in the form of very thin films with desired properties on textile substrates is under development. Hence, the deposition of plasma coatings using combined deposition/etching/sputtering processes enabled the formation of multifunctional surfaces. Multifunctional textile surfaces can thus be obtained by selecting and adjusting the gases used in the plasma chamber to achieve the combined properties such as wettability, functional group density as well as anti-bacterial and bio-responsive surfaces. An industrial scale-up of the process is not far as continuous deposition of such nanostructured plasma coatings on web and fiber has already been demonstrated.

Using plasma co-sputtering of a silver target, Ag nano particles can be in situ embedded within the growing plasma polymer yielding a well-defined size and distribution of nano particles at the coating surface. A homogeneous distribution of Ag nano particles present at the coating surface can be obtained. Hence, an anti-microbial activity has also been achieved. Plasma treatments with or without liquid atomization will become very useful for the textile industry because it can easily modify the textile surfaces with a minimum amount of chemical products [52]. An eco-friendly substitution of traditional padding method to

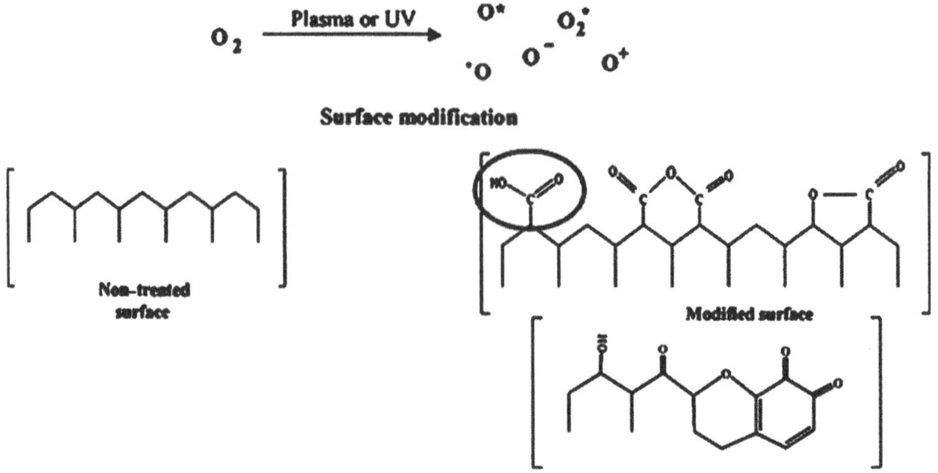

Figure 8. Diagram indicating the modification of cotton textiles induced by plasma or UV pretreatment [29]. Reprinted from A. Bozzi, T. Yuranova, I. Guasaquillo, D. Laub and J. Kiwi, J. Photochem. Photobiol. A: Chem. 174 (2005) pp. 156–164, with permission from Elsevier.

develop fluorocarbon based hydrophobic method can be the fluorocarbon coating using pulse discharge plasma treatment by injecting a fluoropolymer directly into the plasma dielectric barrier discharge. The adhesion of the fluoropolymer to the woven PET is also greatly enhanced by the air plasma treatment.

Bleached and mercerized cotton textiles can be activated by radio frequency plasma (RF-plasma), microwave plasma (MW-plasma), and vacuum-UV light irradiation. Negatively charged functional groups can be introduced on the textile surface by plasma or UV treatment (Figure 8). The UV activated negatively charged surface of textile has been used to anchor $TiO_2$ nanoparticles. This $TiO_2$ deposited textile shows a self-cleaning property i.e. it can easily discolor wine and coffee stains under daylight irradiation. MW-plasma activation of the textile surface followed by $TiO_2$ deposition from a titanium tetra-isopropoxide colloid shows the most favorable discoloration kinetics in the case of the mercerized cotton samples. This is a promising process for the total removal of stains containing persistent colored pigments on the cotton fibers [29].

Also, in case of synthetic textiles, treatment with RF-plasma, MW-plasma and UV-irradiation allows the loading of nano $TiO_2$ by wet chemical techniques in the form of transparent coatings. These $TiO_2$ coated textiles show a significant photo-oxidative activity under visible light in air under mild conditions. Their photocatalytic activity allowed almost complete discoloration of coffee and wine stains in an acceptable time [53]. It has been found that the photocatalytic efficiency of $TiO_2$ can be improved using an organic nanoparticle such as carbon [54]. As the carbon nanoparticle lowers down the band gap of $TiO_2$ by 0.2 eV in the nanocomposite film produced in this study, thus the excitation of $TiO_2$ becomes easier and the photocatalytic activity is improved significantly.

### 3.2.3. Sol-gel

Nanocoating the surface of textiles and clothing is the new approach to the production of highly active surfaces to have UV-blocking, antimicrobial and self-cleaning properties with exceptional durability and minimum coat weight. The sol-gel process is a widely used technique to coat surfaces with nanoscale entities and finds application in a variety of areas ranging from catalysis, electronics, biomedical engineering and material science. In this process, inorganic precursors such as a metal salt or organo-metallic molecule undergo hydrolysis and condensation reactions to form a three-dimensional molecular network of inorganic metal oxide nanoparticles. An example of such hydrolysis and condensation reactions of metal alkoxides to form larger metal oxide molecules by sol-gel process is:

Hydrolysis:

$$M(OR)_4 + H_2O \rightarrow HO - M(OR)_3 + ROH \rightarrow M(OH)_4 + 4ROH$$

Condensation:

$$(OR)_3M - OH + HO - M(OR)_3 \rightarrow (RO)_3M - O - M(OR)_3 + H_2O$$
$$(OR)_3M - OH + RO - M(OR)_3 \rightarrow (OR)_3M - O - M(OR)_3 + ROH$$

Here, M is a metal and R is the alkyl group [55].

Conventional grinding can not produce very small particles (20 to 40 nm) though it is easy to produce by sol-gel processing. When more complex shapes as a fiber or fabric require coating, many of the nanocoating techniques have serious drawbacks and in this case the sol-gel process is attractive because it can be carried out in solution. Coating in solution is generally performed using either precursor molecules or preformed particles to form the layer. Electrostatic interactions, hydrogen bonding, and covalent bonding are some of the associating forces between the coating and the material being coated.

Generally dip coating method is used for sol-gel nanocoating [56]; however, convective assembly system offers higher precision and less cost [57]. Low Temperature Sol-gel based nanocoatings have been recently used to deposit layers of metal oxide nanoparticles on textile fabric surfaces to impact specific functionality, i.e. nanotitania ($TiO_2$) for photocatalytic activity [29]. The cotton surface is first activated by using plasma and UV irradiation before $TiO_2$ colloidal solution is applied to develop sol-gel nanocoating on the fabric surface. This ensures good adhesion of nanotitania particles. It has been found that suitable chemical spacers like polycarboxylic acids can also impart long term stable performance of photocatalytic activity of $TiO_2$ on cotton [58].

The major challenges in sol-gel nanocoatings are controlling the adhesion of the inorganic coatings to the substrates and the uniformity of coating thickness. The homogeneous layer of the inorganic material through nanocoating without the formation of excess inorganic particles in solution is a subject of research. Considerable effort goes into controlling the adhesion of the inorganic coating via the reaction conditions, thereby optimizing the coating quality while preventing disruption or aggregation and preventing the presence of non-coating inorganic material. The majority of the coating procedures have formed monodisperse colloid particles or consistent morphology. The control of coating thickness is of utmost importance so that it can be finely tuned when applications are to be considered. Hence, reaction conditions need to be carefully monitored and well-documented to allow reproducible coatings. Much of this work is already in progress and in the not too distant future we will see numerous nanocoated materials based on sol-gel technology achieving exciting results in different areas of application including textiles.

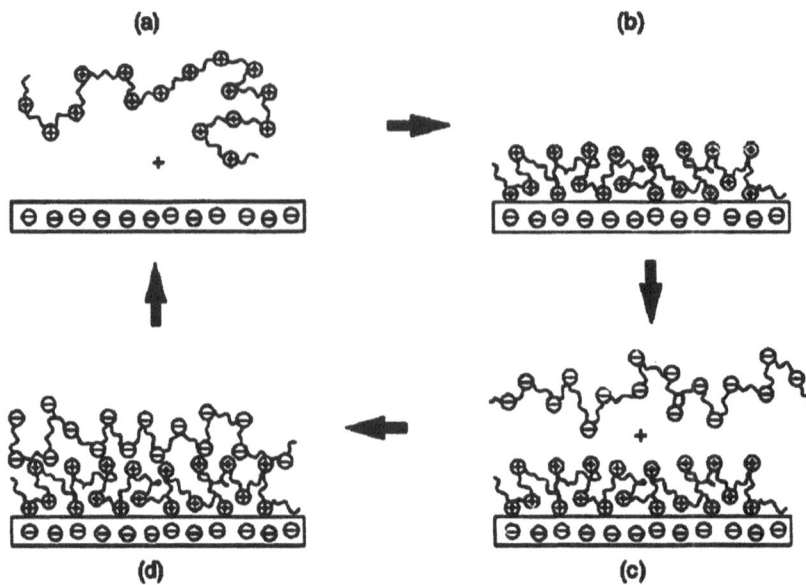

Figure 9. Schematic of L-b-L nanocoatings through self-assembly [18]. Reprinted from M. Joshi, *The impact of nanotechnology in polyesters and polyamides*, in *Polyesters and Polyamides*, B.L. Deopura, R. Alagirusamy, M. Joshi and B. Gupta, eds., Woodhead, Cambridge, UK, 2008, pp. 354–415, with permission from Woodhead Publishing Ltd., Cambridge, UK (www.woodheadpublishing.com).

### 3.2.4. Layer-by-layer

Layer-by-layer (L-b-L) nanocoating technique can be utilized to fabricate thin film coatings, with molecular level control over film thickness and chemistry. No special apparatus is required for the L-b-L process and nanocoatings can be prepared under mild physicochemical conditions. A coating of this type can be applied to any surface amenable to the water based layer-by-layer adsorption process used to construct poly-electrolyte multi-layers including inside surfaces of the complex objects [59]. L-b-L assembly process involves sequential adsorption of oppositely charged polyelectrolytes on a charged solid support that results in multi-layer coatings (or film) of nanometer thick layers (Figure 9).

Systematic modification of the surface of lignocellulosic fibers has been performed by L-b-L nanocoating process to produce negative and positively charged fibers. The fibers are coated with 20–50 nm thick polymer surface layers, which increased the interaction between the fibers during paper formation and helped in obtaining stronger paper from virgin fibers [46]. Cotton fibers offer unique challenges to the deposition of nanolayers because of a unique cross-section as well as chemical and physical heterogeneity of its surfaces. Cationic cotton surface has been successfully coated with alternate layers of anionic and cationic polyelectrolytes, i.e. poly (sodium 4-styrene sulphonate) and poly (allylamine hydrochloride) using L-b-L technique [48]. A study by W. Ali, S. Rajendran and M. Joshi [59] reports that the multilayer formation of polyelectrolytes on cotton surface is sensitive to different process parameters such as pH, temperature, concentration of polyelectrolyte solution, dipping time and addition of salt. Layer by layer technique can also be utilized to create multifunctional textile surface as antifouling, self-cleaning and water resistant coatings for micro-fluids channels and bio sensors. A stable lotus leaf structure has been mimicked to create super hydrophobic surfaces [60]. Antimicrobial silver nanoparticles can be immobilized on nylon and silk fibers by this method. The sequential dipping of nylon or silk fibers in dilute solutions of poly(diallyldimethylammonium chloride) and silver nanoparticles capped with poly(methacrylic acid) lead to the formation of a colored thin film possessing antimicrobial properties. The amount of deposition on both silk and nylon fibers increases as a function of the number of deposited layers though the L-b-L coating on the nylon fibers is not as uniform as on the silk fibers. The deposition of bilayers onto the fibers results in significant bacteria reduction for the silk and the nylon fiber [47]. New antimicrobial synthetic or natural fibers can be designed through this technique.

In a study by M. Joshi et al., L-b-L nanocoating has been carried out on cotton fabric using chitosan as the cationic polyelectrolyte and poly sodium-4-styrene sulfonate as the anionic polyelectrolyte. The process is assisted with ultrasonic treatment for uniform very thin (few nm) deposition of the bi-layers. Thus the produced fabric has good antimicrobial properties; however, the feel, flexibility and breathability of the fabric are not affected [61].

### 3.2.5. Other nanocoating techniques

Several other approaches such as cathode arc vapor technique, electroless coating, in situ polymerization and chemical vapor deposition can also be successfully used to produce nanocoating on textile surfaces.

Metals vaporizing in cathode of a deposition chamber can produce nanoscale multilayer structures having high hardness and better wear resistance. The process is known as cathode arc vapor technique. The high vacuum and high voltage required for this process is often not suitable for textile coating purpose. Considering this, the electroless coating technique is preferable where metal salts chemically dissociate under suitable conditions and metals are deposited on the fabric or fiber. A fine layer of copper coated on polyester fabric through this technique develops conductivity and high quality electromagnetic interference shielding textiles [62]. But the process reduces the tensile strength and increases the drape stiffness of the fabric.

The strength and stiffness issues of base fabric after metal coating shift the interest towards conducting and semi-conducting polymers. Conducting polymers like polypyrrole have the advantage of their reactivity with a number of hazardous gases which leads to a change in conductivity [63]. This property can be utilized to produce gas sensors. Limitations of processability, poor environmental stability, low mechanical strength, poor flexibility and high cost have prevented conducting polymers from making a significant commercial impact [64]. But many of these limitations can be successfully overcome by adhering conducting polymers to strong, flexible textile substrates. In situ polymerization and chemical vapor deposition are the two most suitable techniques to deposit conducting polymers on a fabric or fiber surface.

Monomers like Pyrrole and aniline can be polymerized after coating on to a wide variety of fabrics or fibers by in situ polymerization. The surface resistivity goes down with increase of polypyrrole coating thickness and thus conductivity in textiles can be imparted [65]. In situ polymerized polypyrrole coating on wool fiber has a reasonable stability and dry cleaning fastness [66]. Polypyrrole coated nylon-lycra textiles are used to produce wearable biomechanical sensors which function ideally in a range of applications to monitor human motion [67] and due to their flexibility, they conform to the shape of the human body.

Another suitable process for producing electro-conductive composites is chemical vapor phase deposition of polypyrrole. Comparative morphological and structural analysis on the conducting viscose prepared with both vapor and liquid phase processes show a highly uniform polypyrrole coating on the fiber surface and its partial penetration inside the amorphous zones of the fiber bulk in vapor process. Thus vapor process polypyrrole–cellulose conducting composite textiles show a good performance to light exposure and washing fastness tests [68].

Applications of these electrically conducting, polypyrrole coated fabrics include microwave attenuation, static charge dissipation, and EMI shielding [69]. The microwave absorption characteristics make them useful for military applications such as camouflage and stealth technology. These conducting fabrics are marketed by Milliken & Company under the trademark Contex@. These fabrics are ideal for use as radar absorbing materials. Polypyrrole coated polyester net, having a relatively flat attenuation response over a wide frequency range and overall stability of the microwave absorbance properties, demonstrates the effectiveness of these materials in lightweight camouflage netting applications [70]. The prospect to produce cooling fabric can also be achieved in conducting polymer coated textiles by utilizing the thermoelectric effect [71].

Not only complex conducting polymers, but conductivity of carbon can also be utilized to fabricate yarn-based sensors. Carbon coated piezo-resistive fibers are twisted into yarns and it has been found that the yarn-based sensor can track the respiratory signals more precisely than the fabric based sensors [10]. Thus, smart flexible textile structured sensors can be developed by using conventional textile substrate (fabrics or yarns) coated with a very thin layer of piezo-resistive materials, like polypyrrole or a mixture of rubbers and carbons.

### 3.3. Nanocomposite coating

Polymeric materials coated or laminated on the surface of textiles are an attractive candidate for several high performance applications including defence, biomedical, medical substrates, protective clothing, flexible membranes for civil structures, airbags, geotextiles, agrotextiles and industrial fabrics [35,44,72–74]. The major classes of polymeric materials used to coat a fabric are: natural or synthetic rubber, silicone rubber, PVC or poly-acrylics, teflon, and polyurethanes (PUs) [72,73–77]. All the classes of coating polymers have some basic advantages in terms of functional performance. However, any of the classes alone is not capable to fulfill functional requirements as well as basic coating requirements such as strength, adhesion, wear and coat weight [78]. Combination of more than one of these classes such as teflon/PU [79], silicone modified organics (SMOs) [80], poly (siloxane-urethane) [81], urethane/acrylic composite emulsions [82], fluorinated silicon [83], etc. offer much superior and better performed coatings. Moreover, the fillers are used in coating polymers not only for coloration, fire retardancy, radiation absorption and heat stabilization [72], but also to maximize the durability, abrasion resistance and strength. In some cases, micro encapsulated materials are incorporated into the coating to impart certain functional properties. Microencapsulated phase change materials on textile coating develop stimuli sensitive properties in different applications [11].

With the development of nano-materials, a new horizon is opened up for the scientists to develop nanomaterial dispersed polymeric media for coating on the surface of fabric with satisfactory or even better property and durability. A nanocomposite is comprised of a combination of two or more different substances of nanometer size, thereby producing a material that generally has enhanced specific properties due to the combined properties and/or structuring effects of the components. Enhancement of gas barrier, conducting, antistatic and antibacterial properties with the same or less coat weight is the main concern for the people working in this field. The properties hitherto unknown through coating or by any means can be introduced by nano-technology. The shortcoming of the coating especially in high performance applications can be easily overcome by successful and judicious

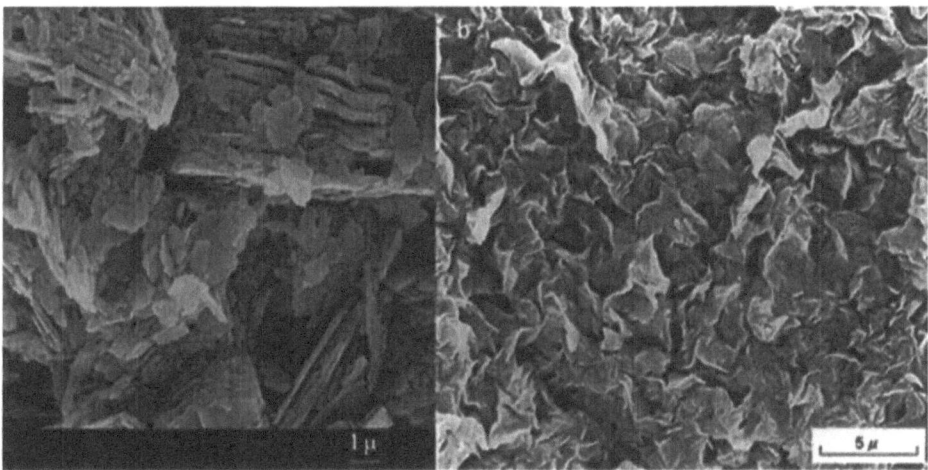

Figure 10. SEM images of clay platelets: (a) kaolin plates and (b) sodium MMT [85]. Reprinted from H.H. Murray, Appl. Clay Sci. 17 (2000) pp. 207–221, with permission from Elsevier.

application of nanomaterials in the coating solution. The nanofiller dimension greatly influences the nanocomposite morphology and polymer coating quality [84]. However, a wide variety of nanofillers, whiskers and nanofibers with structural modification can be used in nanocomposite coatings [77]. Some specific examples of different nanocomposite coatings used in textiles are described here.

### 3.3.1. Clay based nanocomposite coating

Layered silicates dispersed as a reinforcing phase in an engineering polymer matrix are one of the most important forms of "hybrid organic–inorganic nanocomposites". Although several clay minerals are available like kaolinite, smectite, palygorskite and sepiolite (Figure 10), the smectites or bentonites are mostly used because of their high aspect ratio and high cation exchange capacity. Bentonite is the whole rock term for rocks whose dominant clay mineral is smectite. Smectite is a family name, which includes sodium and calcium montmorillonites (MMTs) [85]. Generally, used MMT clay as nanofiller is referred to as nanoclay, layered silicate, 2:1 nanoclay etc.

Layered silicates particularly MMT clays both natural as well as synthetic (alternatively referred to as 2:1 layered aluminosilicates, phyllosilicates, clay minerals and smectites) are the most commonly used inorganic nano elements in polymer nanocomposite research to date. The chemical structure is:

$$M_x (Al_{4-x} Mg_x) Si_8 O_{20} (OH)_4$$

M = monovalent cation; x = degree of isomorphous substitution (between 0.5 and 1.3).

Their crystal lattice consists of two-dimensional layers where a central octahedral sheet of alumina or magnesia is fused to two external silicate layers. Depending on the precise chemical composition of the clay, the sheets bear a charge on the surface and edges, this charge being balanced by counter ions, which reside in part in the interlayer spacing of the clay. Stacking of the layers by weak dipolar or Van der Waal forces lead to interlayer or galleries between the layers. The galleries are normally occupied by inorganic cations ($Na^+$,

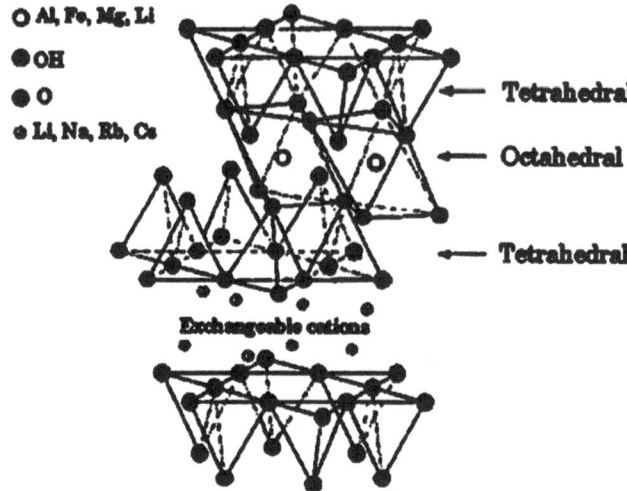

Figure 11. Structure of 2:1 layered silicate [18]. Reprinted from M. Joshi, *The impact of nanotech-nology in polyesters and polyamides*, in *Polyesters and Polyamides*, B.L. Deopura, R. Alagirusamy, M. Joshi and B. Gupta, eds., Woodhead, Cambridge, UK, 2008, pp. 354–415, with permission from Woodhead Publishing Ltd., Cambridge, UK (www.woodheadpublishing.com).

$Ca^{++}$) balancing the charge of the oxide layers. These cations are readily ion-exchanged with a wide variety of positively charged species. The number of exchangeable interlayer cations are also referred to as the cation exchange capacity (CEC). This is generally expressed as milliequivalents/100 g and ranges between 60 and 120 for relevant smectites. In layered silicates, the Van der Waals interlayer or gallery containing charge compensating cations ($M^+$) separates covalently bonded oxide layers, 0.96 nm thick, formed by fusing two silica tetrahedral sheets with an edge shared octahedral sheet of either alumina or magnesia (Figure 11).

Pristine layered silicates usually contain hydrated $Na^+$ or $K^+$ ions. One important consequence of the charged nature of the clays is that they are generally highly hydrophilic species and therefore naturally incompatible with a wide range of polymer types. Therefore clay often must be chemically treated in order to make it organophilic. When the inorganic cations are exchanged by the organic cations, these are called organically modified layered silicates (OMLS). Generally this can be done by ion exchange reactions with cationic surfactants including primary, secondary, tertiary and quaternary alkyl ammonium or alkyl phosphonium cations. These cations in the organosilicate lower the surface energy of the inorganic host and improve wetting characteristics of the polymer matrix or in some cases initiate the polymerization of monomers to improve the strength of the interface between the inorganic and polymer matrix.

Due to their high aspect ratio, MMT clays are ideal for reinforcement. Highly exfo-liated morphology of the clay enhances the thermal stability, mechanical properties and anti-corrosion protection in polymeric nanocomposites [86]. But the nanolayers are not easily dispersed in most polymers due to their preferred face-to-face stacking in agglom-erated tactoids. Dispersion of the tactoids into discrete monolayer is further hindered by the intrinsic incompatibility of hydrophilic layered silicates and hydrophobic engineering plastics. Though clays are hydrophilic, formation of nanobubbles at the edge of hydrophilic clay may attribute its flocculation and attraction to hydrophobic surfaces [87]. In order to make it dispersible in organic solvents or in polymers, high ion exchange capacity of

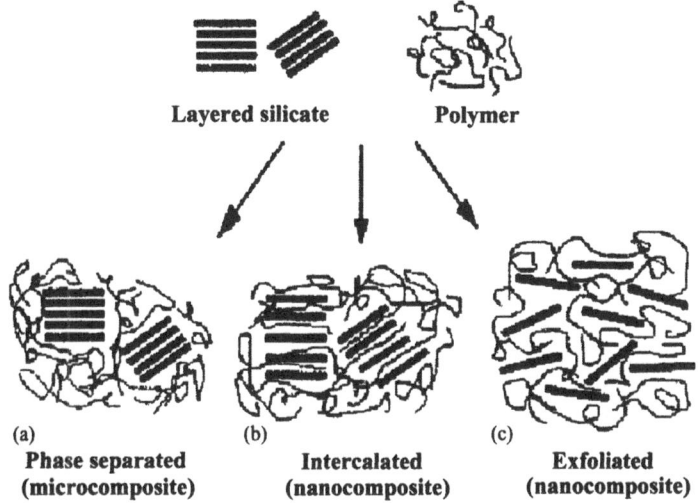

Figure 12. Different types of composite from the interaction of layered silicates and polymers: (a) phase separated microcomposite; (b) intercalated nanocomposite and (c) exfoliated nanocomposite [89]. Reprinted from M. Alexandre and P. Dubois, Mater. Sci. Eng. 28 (2000) pp. 1–63, with permission from Elsevier.

MMT clays is usually utilized [88] and organic long chain molecules are inserted into the clay gallery. Different approaches have been initiated to successfully disperse organically modified clay in polymer matrix so that its full potential can be utilized [89,90]. Incomplete exfoliation and disoriented platelets lead to reduction in reinforcement efficiency. The same clay filler can form microcomposite or nanocomposite with polymers depending on its degree of intercalation and exfoliation (Figure 12). The dispersion of organophilic MMT in organic solvents, monomers or prepolymers for polymer coatings are characterized by clay interlayer distance (d-spacing), the rheological behavior of clay suspensions and the macroscopic swelling or the final suspension morphologies [91].

The first spectacular industrial commercial application of polymer nanocomposites with clay minerals as nanofillers was in the 1990s from Toyota Central Research Laboratories [92]. After this other commercial polymer nanocomposites with clay nanofillers have been regularly documented for four main properties e.g. mechanical (automotive industry, etc.), barrier (packaging, film and bottle industry, etc.), fire retardancy (cable industry, etc.) and physical and optical (electronic industry, batteries, proton exchange membranes fuel cells) [89]. Their applications also include their use as core in tennis balls, biomaterials, and particularly biomimetic materials [93,94]. Nanocomposite coatings based on Na-MMT clay show better corrosion resistance and barrier properties [95]. The plate like structure of clay reinforces the polymer under tensile load while the improvement is much less under compressive force [96]. The other advantage of nanoclay is that the polymers can bear more load at elevated temperature when reinforced with it [97].

Suitable modification or use of some coupling agent with nanoclay always shows better mixing, dispersion and property enhancements in polymer matrix. For example it has been observed that the silane modified clay dispersed in PU gives better physical, mechanical and thermal properties than commercial clay-based nanocomposites [98,99]. This can be applied as a coating to have better property realization. Hydrogen, helium, oxygen and water permeability of nylon-6 film significantly decreases with MMT clay loading [100]. With the increase in exfoliated vermiculite clay loading the gas barrier property of butyl rubber is reported to increase during coating application [101]. PU/clay nanocomposite coating has superior gas barrier property than the pure PU coating. When coated on nylon fabric for inflatables it reduces the gas permeability significantly [102]. A homogeneous dispersion of surfactant modified clay is reported in acrylonitrile matrix [103]. Subsequent polymerization leads to highly exfoliated nanocomposite. This process has been used to produce acrylic nanocomposite coating. Nanoclays also increase the fire retardancy of clay/acrylic nanocomposite coating by formation of a ceramic like protective barrier on charring [104]. Moreover, study on nylon-6 shows that the nanoclay can enhance the flame retardancy when used with other flame retardants but nanoclay itself has no significant enhancement. This may be due to poor thermal stability of the modifier used during preparation of modified nanoclay [105]. One interesting observation is when applied together, nanokaolin and nano-HAO (Hydroxyl aluminum oxalate) produce synergistic effect on flame retardancy [106].

In general, use of a small amount of organoclay increases the thermal stability and mechanical properties and reduce the gas permeability in polymeric nanocomposites [107–114]. Nanoclays have been added in a variety of polymeric coating formulations such as PU, acrylic, butyl rubber, polyimide and epoxy to impart some of these properties through nanocomposite coating. To control the mechanical properties of a nanocomposite coating the process route and surface modification of nanoclays are most critical [115]. Type of surfactant used to modify clay structure plays a dominant role in determining the properties of PU/clay nanocomposites [116]. Organically modified reactive clay incorporated during bulk polymerization of thermoplastic PU substantially increases the mechanical properties, thermal stability and abrasion resistance [117,118]. Aqueous emulsion of PU ionomers, reinforced with organoclay produce nanocomposites where the organoclay is effectively intercalated or exfoliated so that the thermal and water resistance are enhanced while transparency is marginally reduced than the pristine polymer [119]. In PU based shape memory polymers the addition of clay increases the shape recovery stress which can be utilized to enhance the reliability and robustness of the shape memory polymeric coatings [120].

The anticorrosive behavior of polyimide coating is found to be greatly improved with very low loading of MMT clay [121]. Same observations are made with epoxy-clay nanocomposite coatings [122]. Thus the large amount of ongoing researches indicates the commercial viability of nanoclay based polymeric products in the near future.

### 3.3.2. Silica based nanocomposite coating

Nanosilica is the most promising nanoparticle to create nano-roughening on textile surface and thus produce super-hydrophobic surfaces successfully. Nanocomposite coating of silica nanoparticle in perfluoroacrylate based water repellent agent produces super-hydrophobic cotton fabrics. The combined treatment of nano-roughening and surface tension lowering agent reduces the amount of non eco-friendly fluoro compound to 0.1% [123]. As such silica can not bind to cotton surface. So the nanoroughening imparted by silica nanoparticles are not durable. However, silica nanoparticles can be suitably functionalized to adhere with cotton surface. Amino functionalized silica nanoparticles applied on epoxy functionalized cotton fabric shows excellent super-hydrophobicity ($\sim$170°). The functionalization of silica and cotton surface helps to create a strong bond between the fiber and particles. Thus, produced nano roughening on fiber surface is very robust and durable [124]. Besides the water repellent lotus effect, nanosilica is used to produce wrinkle resistance in silk [125], to reduce the friction and wear of nylon-6 [126] and PU coating [127]. As nanosilica has very high thermal stability, it also reduces the peak heat release of polymer matrix [128], thus making the fabric useful for flame retardant textiles.

### 3.3.3. Carbon nanomaterial based nanocomposite coating

Carbon nanomaterials are the most researched area for high performance nanocomposite coatings due to their unique conducting, mechanical and responsive properties. Among them carbon nanotube (CNT), both single (SW) and multi-wall (MW), is listed as number one for its exceptional mechanical properties, very high electrical as well as thermal conductivity and electroactivity [129]. Thus, the polymer nanocomposites based on CNT has tremendous potential for high performance textile applications.

Carbon nanotubes (CNTs) were first reported by Ijima in 1991. Since their discovery, CNTs have been the focus of considerable research because of the unique and unprecedented mechanical, electrical and thermal properties. Basically, CNTs are long, slender fullerenes where the walls of the tube are hexagonal carbon (graphite structure) and often capped at each end by hemi – fullerenes. CNTs can be visualized as a sheet of the graphite that is rolled into a tube. Unlike diamond, where a 3-D diamond cubic crystal structure is formed with each carbon ($sp^3$ hybridized) atom having four nearest neighbors arranged in a tetrahedron, graphite is formed as a 2-D sheet of carbon ($sp^2$ hybridized) atom arranged in a hexagonal array. In this case each of the carbon atoms has three nearest neighbors.

CNTs are of two kinds – single walled (SWCNT) and multi walled (MWCNT). SWC-NTs can be considered as a single graphene sheet rolled up into a seamless cylinder where graphene is a monolayer of $sp^2$ bonded carbon atoms. MWCNTs are multilayered and simply composed of concentrically arranged single walled CNTs with a central hollow core with interlayer separation of almost 0.34 nm, an indication of the interplane spacing of graphite. A special case of MWCNTs is double walled CNTs (DWCNTs) that consist of two concentric graphite cylinders. DWCNTs are expected to exhibit higher flexural modulus than SWCNTs due to two walls and higher toughness than regular SWCNTs due to their smaller size.

The properties of nanotubes depend on atomic arrangement or how the sheets are rolled, the diameter and length of the tubes and the morphology of the nano structure. The graphene sheets can be rolled up into tubes in various ways and are described by the tube chirality (or helical or wrapping), which is defined by the chiral vector ($C_n$) and the chiral angle ($\theta$).

$$C_n = n_a + m_a,$$

where the integers (n, m) are the numbers of steps along the unit vectors ($a_1$ & $a_2$) of the hexagonal lattice. Using m & n, the three different types of orientation of the carbon atom around the nanotube circumference is specified as armchair, (n = m), zigzag (n = 0, m = 0) or chiral (all others) (Figure 13).

The chirality of the CNTs has a significant effect on its properties, especially on the electronic properties. All armchair SWCNTs are metallic with a band gap of 0 eV. SWCNT with n-m = 3i (i being an integer $\neq$ 0) are semi metallic with a band gap of few mev, while SWCNT with n-m /= 3i are semi conductors with a band gap of 0.5–1 eV. MWCNTs contain a variety of tube chirality's, so that their properties are more complicated to predict.

MWCNTs and SWCNTs are mainly produced by three techniques: arc discharge, laser ablation and chemical vapor deposition (CVD). The various aspects of nanotube production, purification, suspension, filling, functionalization and application as well as the fabrication

Figure 13. Schematic diagram showing how a hexagonal sheet of graphene is rolled to form CNTs of different chirality [18]. Reprinted from M. Joshi, *The impact of nanotechnology in polyesters and polyamides*, in *Polyesters and Polyamides*, B.L. Deopura, R. Alagirusamy, M. Joshi and B. Gupta, eds., Woodhead, Cambridge, UK, 2008, pp. 354–415, with permission from Woodhead Publishing Ltd., Cambridge, UK (www.woodheadpublishing.com).

and characterization of polymer nanocomposites with various types of nanotubes have been extensively reported in literature and are not being discussed here. All known preparations of the CNTs give mixtures of nanotube chirality's, diameters and lengths along with different amounts and types of impurities. This CNT heterogeneity makes purifications of CNTs an essential requirement before they can be put into any application including synthesis of polymer /CNT nanocomposites.

CNTs exhibit exceptional material properties that are a consequence of their symmetric cage like structure and exceed those of any previously existing materials. The high aspect ratio of CNTs coupled with strong intrinsic Van der Waals forces of attraction between nanotubes combine to produce ropes and bundles of CNT. SWCNTs where the attractive force is very high (0.5 eV per nm of CNT to CNT contact) are more prone to form ropes and bundles. Ropes refer to collections of SWNTs and are more uniform in diameter to form a hexagonal lattice, while bundles are non-crystalline collections of SWCNTs or MWCNTs. It is for this reason that dispersion of CNTs in solvent to produce suspensions, need assistance of either sonication or surfactant addition or both combined besides mechanical stirring.

The dispersion of untreated CNT in polymeric media is very difficult due to the intense Van der Waals force holding them together. This is the main deadlock to achieve CNT's potential as nanofiller. The tremendous properties of raw CNT have not been realized in polymeric nanocomposite form until today for this reason. However, acid treatment leads to the partial destruction of the nanotube wall and hydrophilic groups are introduced into the tube structure which results in easier dispersion and better mixing. It has been observed that the wear properties of PU coating are greatly improved after acid functionalization of carbon fiber [130]. Similarly, when CNT is functionalized, it is dispersed and attached better with the polymer so their multifunctional properties can be utilized. The functionalization can be done using strong acids or ozone treatment under UV light. Silanes can act as a coupling agent between the reduced CNT and the polymer [131]. Acid functionalized CNTs can also form urethane linkage with diisocyanate (Figure 14) and thus produced surface modified carbon nanotube reinforced PU composite coating which has high coefficient of

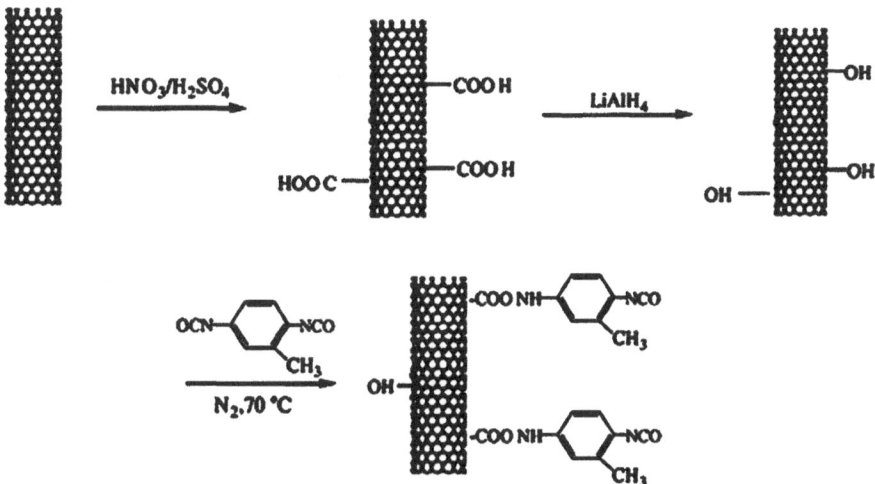

Figure 14. Functionalization of multiwall CNTs and addition with toluene-2,4-diisocyanate [132]. Reprinted from H.J. Song, Z.Z. Zhang and X.H. Men, Eur. Polym. J. 43 (2007) pp. 4092–4102, with permission from Elsevier.

friction and wear resistance [132]. One interesting thing is that the coefficient of friction is decreased with increasing sliding speed and applied load.

The problems with rolled graphene i.e. CNT and carbon nanofiber switch some interest towards layered graphene or nanographite. Nanographite/nanographene is nanometer sized fragments of graphite where the stack thickness is essentially in nanometer size but the length and width of the flake is in micron size. This is a promising carbon nanoparticle that can be used in nanocomposite coatings based on PU and epoxy. Nanographite has unconventional nanoscopic electronic and magnetic features due to the presence of non-bonding open edges with $\pi$ electron state which is entirely different from bulk graphite [133]. Chemical vapor deposition may be the technique to successfully coat nanographite while the electro and magneto conductivity of such film depends on the temperature and orientation of the graphene layers [134] and also the geometric structure of the graphite layer [135]. Intercalation of nanographite or stacked nano-graphene sheets is of particular interest since the electronic structure of the intercalated nanographite is expected to impart features different from those of bulk counterpart in relation to the presence of edge states. Nano-graphite can be intercalated with potassium, fluorine, bromine or iodine [136].

The iron-nickel co-deposited nanographite is found to be a good microwave absorbent. The PU based nanocomposite film prepared with a number of fillers shows that the nanographite is capable of better microwave absorption in comparison to graphite and the metal coating significantly enhances its microwave absorption in radar frequency (8 – 18 GHz) at 10% filler loading. This study has the potential to produce flexible radar absorbent coatings on textiles for defense applications [137]. Incorporation of expandable graphite enhances the fire retardancy of PU without affecting the thermo-oxidative behavior of PU coating [138]. The gas permeability of textile structured nanographite/epoxy composite shows very little increase on cryogenic cycling [139]. Nano-graphite intercalated compounds are more sensitive to helium gas adsorption than bulk-graphite compounds [140]. Thus low temperature gas permeability based sensor application with adequate robustness is possible using expandable graphite coatings.

### 3.3.4. Other nanoparticles based nanocomposite coating

Several other nanoparticles are also used in nanocomposite coatings for performance enhancement and to overcome the limitations of traditional polymers used for coating. Metals, metal oxides and hybrid nanoparticles are usually incorporated into coatings to impart different functional properties.

One such example is encapsulated di-ammonium hydrogen phosphate (DAHP) in PU shell to improve the flame retardancy of the PU-urea coating of cotton or cotton-polyester blended fabrics. The expected advantages of this new concept of encapsulated fire retardant agent lie in its being compatible with the polymeric matrix in order to give a permanent fire retardant effect and while being an efficient fire retardant intumescent formulation for many materials. Microcapsules and their components (e.g. DAHP and PU shell alone) have been examined by thermal analysis in air. The experimental and theoretical thermo gravimetric responses of weight loss for the two types of microcapsules show evidence of interactions between DAHP core and PU shell. The two types of microcapsules exhibit a significant stabilization at high temperatures (above 450°C) with production of thermal stable char that can be attributed to the development of an intumescent structure. The reaction to fire of cotton fabrics coated by fire retardant polyurea loaded with neat and microencapsulated DAHP was studied and it has been observed that the coatings with microcapsules show decreases in propensity of flame with regard to virgin polyurea coating. The coatings with microcapsules evolve smaller quantities of smoke and CO than virgin polyurea and all other FR coatings. So, in comparison with the conventional PU-phosphate combinations, this encapsulated flame retardant agent is a better choice because phosphate in this form is water insoluble and thus gives a permanent flame retardant effect [141].

UV absorbent zinc oxide and $TiO_2$ nanoparticle-embedded acrylic coatings have been developed to protect base fabrics from sun-induced degradation. Nanoparticle-embedded acrylic coating on aramid Kevlar fabric largely prevents degradation of Kevlar fabric after a week of UV exposure [142]. Nanoparticle-embedded acrylic coatings can absorb a sufficient amount of UV radiation yet scatter little and these have been developed to protect base fabrics from sun-induced degradation. Zinc oxide and $TiO_2$ nanoparticles with diameters ranging from 15 to 70 nm (5 wt%) were dispersed in acrylic emulsions. Nanoparticle-embedded acrylic films of 10 mm and 20 mm thick were prepared and bonded to Kevlar fabric. Mechanical tests as well as infrared, visible and UV spectroscopy were used to characterize nanoparticle-embedded acrylic emulsions and coated Kevlar fabric. The changes in mechanical and chemical properties of Kevlar fabric after a day and week of intense UV exposure were assessed using tear and strip tensile testing, UV, visible and infrared spectroscopy, and wide and small angle X-ray analysis. Tear and tensile data showed that 20 $\mu$m thick nanocomposite coating largely prevented degradation of Kevlar fabric, allowing only 5% of the degradation that occurred in the unprotected fabric after a week of UV exposure. Again, nano layer double hydroxide and nano $TiO_2$ dispersed in acrylic and silicon resin improve the resistance of coating to moisture, ageing, UV and flame [143].

The friction and wear lead to increased service and repair cost. The choice of coating to reduce friction and coating in a product depends on the product use and its lifetime [144]. Use of nanoparticles during coating applications helps to increase the wear properties of the coating. Nano-$TiO_2$ effectively reduces the coefficient of friction and wear properties of nylon-66 especially in case of high pressure and high velocity conditions. Nano-$Al_2O_3$ shows similar behavior [145]. The effect of nanoparticles on the reduction of friction and wear is explained as three bodies (fabric, polymer and nanoparticle) contact instead of direct surface contact, which results in lowering of abrasive force with positive rolling effect of the nanoparticles between the material pairs [146].

Long-term antimicrobial activity can be imparted in many coating formulations through the incorporation of nanomaterials. The gold nanoparticles have been used in PU cardio-vascular biomaterials for higher thermal and biostability and trials on rat show better results over pure PU [147]. The desire for permanent coatings to impart long-term antimicrobial or bacteriostatic properties to coated products has been expressed in a variety of industries, including healthcare, industrial and institutional cleaning, food processing, food service, and general paints and coatings. NanoTek®zinc oxide, copper oxide or doped zinc oxides can be fully dispersed into a wide variety of coating formulations, including urethanes, acrylics and vinyl acetates and have shown utility in UV curable and thermosetting coatings as well as in water-based coating systems [12].

### 3.4. Nanodyeing

Although nanotechnology touches into all aspects of textile chemical processing, a relatively untouched area is dyeing. To develop textile products which are more and more sophisticated and multifunctional to meet growing customer needs and to assure a share in the market, nano-technological advances have to be introduced in the dyeing houses. Nanotechnology can also reduce the use of water as the nanostructure and surface functionality can be imparted on fabric using dry techniques.

Some reports on bulk or surface modification of fibers or fabric have described the improvement or change in the dyeability of the fabric and its fastness properties. A boon of nanotechnology is that it can provide a solution to overcome the most controversial and eco-hazardous effluent problem in a process house. Conventional finishing techniques applied to textiles (dyeing, stain repellence, flame retardant, antibacterial treatments) generally use wet-chemical process steps and produce a lot of wastewater. A composite reactor prepared with nanoTiO$_2$ effectively degrades the effluents including dyes and pigments in waste water [148]. Research is going on to use nanotechnology in water treatment of textile wet processing units.

#### 3.4.1. Modification in substrate and dye

Dyeing itself is difficult for some synthetic fibers such as PP due to its highly crystalline and non-polar nature which provides no sites for dye uptake. The fiber morphology can be altered by introducing nanofillers into such fibers and thus, dyeing behavior of a fiber can be modified. Nanocomposite fibers based on quaternary ammonium modified nanoclay/PP fiber show affinity towards acid and disperse dyes the clay reduce the PP fibers crystallinity which in turn helps to increase the access of disperse dye in the fiber and the cationic dye sites from ammonium iron helps to exhaust acid dye. Thus a dyeable PP can easily be produced [149].

When nanoclay is incorporated in nylon fiber through melt spinning, it has been observed that the crystallization temperature increases and the melting point of nylon decreases [150]. The quantity of amorphous regions is higher in clay filled nylon yarns. Thus the clay filled yarn dyes itself faster with disperse dyes than unfilled nylon yarn. However, the dyeing behavior of such fibers are opposite with acid dyes and metal complex dyes i.e. the dyeing rate and color value is less than pure nylon fiber. This is because the anionic groups of nanoclay interact with the amino sites and dying sites of the fiber are blocked, thus clay prevents the fixation of the acid or metal complex dyes on nylon fiber.

Particle size of pigments plays an important role to achieve high colorfastness in cationic modified silk dyeing. Pigment dyeing is discarded by the industry as the fabric handle becomes harsh. But use of ultrafine nano pigments helps to overcome this limitation [151]. The surface of silk is modified by a cationic reagent, N-(3-chloro-2-hydroxypropyl) trimethylammonium chloride (CHTAC) in order to enable the fiber to be dyed with ultrafine pigments by an exhaust process. The effects of cationization pretreatment conditions such as the amount of cationic reagent, pH, treatment temperature and time on color yield have been discussed in detail. The result shows that the condition suitable for modification treatment is that the concentration of cationic reagent is 10 g/l, pH 8, liquor to material ratio of 100:1 and 60°C for 30 min. The whiteness index decreases with the increase of alkali of cationic treatment. The crock fastness and wash fastness of silk dyed by pigment exhaust dyeing achieved is 3–4 and 4 scale, respectively. The treated silk fabrics still retain a soft handle because the bending rigidity and hysteresis increase slightly after the cationization pretreatment and dyeing procedure. It has been demonstrated that properties of surface modified silk dyed with ultrafine pigment by exhaust process are acceptable. As the particle size of the pigments goes down to nano level, the dyeing becomes easier, exhaustion is higher and above all even the silk fabric feel has not changed much. The fastness issue of the pigment dyeing can also be successfully encountered.

Pigments have some inherent problem in terms of dispersion and durability (UV fastness). In order to improve their weather durability, heat resistance, color strength and dispersion ability encapsulation with polystyrene and silica has been successfully carried out by sol-gel method [152]. Pigments protected under silica shell show improved thermal stability, dispersion and UV shielding property. The layer-by-layer incorporation of nano-silica increases the scattering of the UV rays. Thus the color stability of pigments is enhanced after such encapsulation.

### 3.4.2. Surface texturing using plasma

The nanostructured surfaces have attracted the attention of the researchers as high color value and intensity can be obtained because of the increase in surface area. The nanoparticles are mainly introduced to build nanostructured surfaces through wet-chemical coatings or compounds. But agglomeration and uneven distribution of nanoparticles at the fabric surface can lead to deterioration of fabric properties such as handle, feel and strength. Nanoporous structures can also be obtained through plasma treatment on the textile surface and have the potential to avoid these issues and yield an even higher surface area. Plasma treatment, a dry and eco-friendly technology, is offering an attractive alternative to add new functionalities such as water repellency, long-term hydrophilicity, mechanical, electrical and antibacterial properties as well as biocompatibility due to the nano-scaled modification on textiles and fibers. At the same time, the bulk properties as well as the touch of the textiles remain unaffected. Plasma polymerization can be used to obtain highly cross-linked plasma coatings, e.g. by using simple hydrocarbons. When non-polymerizable gases such as Ar, $CO_2$ or $NH_3$ are added to the gas discharge, chemical etching and sputtering effects can be achieved to modify the film growth. Thus, interconnected voids could be obtained within a thin film. Further nano particles with homogeneous size and spatial distribution can also be embedded in situ by a plasma polymerization/co-sputtering process.

Plasma polymerization of acetylene mixed with ammonia is used for both deposition and etching processes resulting in a nanoporous, cross-linked network with accessible functional groups on textile surface. Mixtures of $C_2H_2/NH_3/Ar$ in low pressure radio frequency (RF) plasma discharge resulted in nanoporous coatings and Ag nanoparticles can also be incorporated from a silver target during such multifunctional porous structure formation [153]. Co-sputtering of a silver target with Ar enabled the in-situ incorporation of Ag nano particles within the functional plasma coating to add anti-microbial properties into it. While using RF discharges with acetylene, the growth mechanism of plasma-polymerized hydrocarbon coatings are modified by admixture of $NH_3$ to induce etching processes during deposition. The ratio of $C_2H_2/NH_3$ controls the deposition and erosion on the surface. Rivaling deposition/etching processes thus support the generation of nanoporous, highly cross-linked plasma coatings containing accessible functional groups. These plasma coatings can be used as permanent hydrophilic treatment or for substrate-independent dyeing when deposited on textile fabrics. Increasing color intensity of acid dye with film thickness proved that accessible amine groups were deposited within an interconnected nanoporous hydrocarbon structure. Beside the advantage of obtaining a substrate-independent dyeing on textile fabrics and fibers, these kinds of coatings were also found to be permanent hydrophilic in nature due to the suitable surface texturing. A complete wetting of textiles over long periods (more than a year) can thus be achieved.

### 3.4.3. Surface texturing using UV/O₃

Continuous surface treatment using an electrode-less UV bulb introduces nanoscale surface roughness. The dyeability of such surface treated polyester fabrics to disperse dyes is chemically similar to that of untreated fabrics. But irradiation significantly increases the depth of shade with black disperse dye up to 8% due to the enhanced surface roughness. The trend remains the same when the UV irradiation is given both after and before dyeing. Color fastness to laundering and rubbing of these dyed polyester fabrics is excellent because of surface-limited treatment of the polyester fabrics.

UV/O₃ treatment produces micro-sized and nano-sized surface roughness by selective photo-degradation. This in turn encourages light to scatter and interfere destructively depending on the height of the surface roughness. The UV/O₃ treatment has been used as a nano-roughening method for color deepening of the polyester fabrics which can substitute current plasma and sputter etching techniques (Figure 15) [154].

## 4. Nanotechnology in multifunctional fibers and nanofibrous forms

Nanotechnology has led to the development of new fibers and nanofibrous materials suitable for a number of high-end applications. Advanced nanocomposite fibers have unprecedented properties. Hydrophilic polyester, dyeable PP and antimicrobial nylon are a few such examples. All of these are only possible because of the nanotechnology. The incorporation of nanofillers into the polymer matrix before it is spun into a fiber is the key to developing hitherto unknown properties of the fiber. Such fibers are known as nanocomposite fibers as the fillers are dispersed in nano-dimension into the fiber-forming polymer. The wide choice of nanoparticles gives the advantage of preparation of a range of multifunctional, smart nanocomposite fibers that are successfully used in various high-performance textile applications.

Another popular class is the nano-fibrous form in which the fiber dimension itself is in submicron or nano range. These are called nanofibers and are produced in the form of nonwoven mesh in which each fiber is submicron size in diameter. The nanofibers have attracted an immense interest in recent years especially due to the renewed interest in the development of electrospinning process. Traditional methods of polymer fiber production, including melt spinning, solution spinning and gel state spinning, rely on mechanical

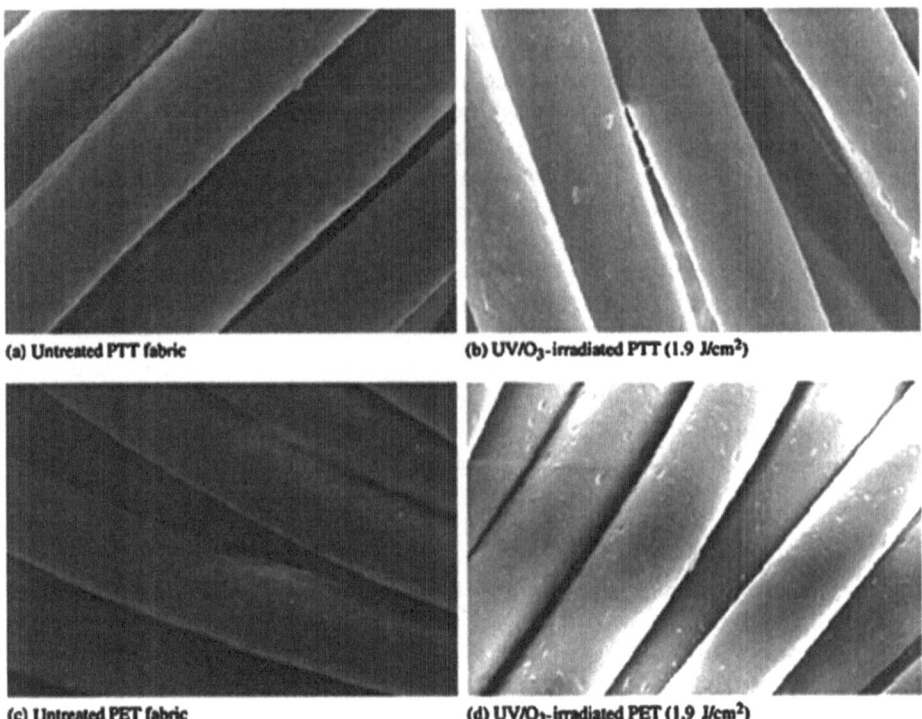

(a) Untreated PTT fabric                    (b) UV/O₃-irradiated PTT (1.9 J/cm²)

(c) Untreated PET fabric                    (d) UV/O₃-irradiated PET (1.9 J/cm²)

Figure 15. SEMs of untreated and UV/O$_3$ irradiated PET and PTT fabrics [154]. Reprinted from J. Jang and Y. Jeong, Dyes Pigm. 69 (2006) pp. 137–143, with permission from Elsevier.

forces to produce fibers by extruding polymer melt or solution through a spinneret and subsequently drawing the resulting filaments as they solidify or coagulate. By using these methods, the consistently producible minimum fiber diameter is in the range of 5–500 microns, that is of the order of a micron, whereas the diameter of polymeric nanofibers is in the submicron range (from 50 to 500 nm). An extraordinarily high surface area per unit mass, high porosity and lightweight are the main characteristics of the nanofibrous webs.

In the subsequent chapters, the development, production process, properties and application of various nanocomposite fibers and nanofibers are described.

### 4.1. Nanocomposite fiber

Nanocomposite fibers are the advanced new class of polymer nanocomposite materials with an ultrafine dispersion of nanoparticles or nanoparticles in a polymeric matrix. When compared with neat polymers or conventional composites, nanocomposite fibers have much superior properties. The properties include high modulus, increased strength, improved heat resistance, decreased gas permeability and flame retardance at very low loadings (<5 wt%) of nanofillers.

On the one hand, in conventional composites, the mechanism of property enhancement follows the rule of mixtures. On the other hand, in nanocomposites, the property enhancement exceeds the theoretically predicted values giving a unique combination of properties synergistically derived from both the nanomaterials and the polymer matrix system. The volume and influence of the interfacial interactions increases exponentially with a decrease in filler size and thus forms an additional separate phase known as interphase, which is distinct from the dispersed and continuous phases and hence influences the composite properties to a much greater extent even at a low nanofiller loading (<5 wt %) [155,156]. Therefore, their properties are much superior to conventional composites. The lighter weight of polymer nanocomposites because of the low filler loadings as compared to conventional composites creates further interest. They are usually transparent as scattering is minimized because of the nanoscale dimension involved and are still processable in many different ways even with nanoscale fillers embedded in the polymer matrix. The major challenges in nanocomposites are however ascertaining a high degree of dispersion of nanomaterials in the polymeric resin during nanocomposite synthesis and processing. Since the advent of Nylon 6/MMT nanocomposites developed by Toyota Motor Co. (Aichi, Japan) [157], relentless efforts are being made globally to successfully extend this concept to almost all types of polymer matrices, with a range of nanomaterials used as reinforcing entities. Polymer nanocomposites thus exploit the fascinating and useful properties of nanomaterials for a variety of structural and non-structural applications such as automotives and packaging industries, building and construction, electrical and electronics, and sports and medical devices [158–160].

The tremendous potential of polymer nanocomposites can be utilized in the form of nanocomposite fiber, as the nanocomposite fibers offer properties that are lacking in the commodity synthetic fibers. Nanocomposite fibers that contain nanoscale embedded rigid particles as reinforcements show improved high-temperature mechanical properties, thermal stability, useful optical, electrical, barrier or other functionalities such as improved dyeabilities, flame retardance, antimicrobial properties and so on. These novel two-phase fibers in which the dispersed phase is of nanoscale dimension will make a major impact in tyre reinforcement, electro-optical devices and other applications such as medical textiles, protective clothing and so on [161].

The work on spinning of polymer nanocomposites started about 10 years ago, and several research groups across the world are exploring the synthesis, fiber processing, structure-property characterization and correlation and molecular modeling of these unique new nanocomposite fibers. Generally, the nanocomposite fibers have the dimension of conventional fibers, that is in the micron range, but the nanofillers are dispersed uniformly into their polymer matrix. The nanofillers may have only one dimension in the nano range – lamellar (e.g. layered silicate clays), two dimensions in the nano range – fibrillar (e.g. CNTs) or all three dimensions in the nano range – spherical (e.g. metals/metal oxide nanoparticles and POSS [polyhedral oligomeric silsisquioxanes]). All the major types of nanofillers have been incorporated into polymeric nanocomposite fibers such as layered silicates (MMT), CNTs and nanofibers, metal oxide nanoparticles ($TiO_2$, $ZnO$, $SiO_2$ and so on) and hybrid nanostructured materials (POSS). A brief review of the work being done in the area of nanocomposite fibers is described in the following sections.

### 4.1.1. Polymer/clay nanocomposite fibers

Incorporation of organoclay into polymer matrix modifies the polymer's performance as a fiber during various textile applications. The main effects are observed in physical properties such as enhanced tensile modulus and strength, reduced thermal shrinkage, controlled electrostatic behavior, high storage and reduced loss modulus under dynamic mechanical testing conditions. Along with these properties, presence of hydrophilic clay also enhances moisture absorption (at a slow rate), dyeability, biodegradability and chemical resistance. Clay-based nanocomposite fibers exhibit improved weatherability, as clay platelets block

UV irradiation. The inherent nature of clay to reduce burning rate and form char results in reduced flammability of these fibers. With these improved properties, these fibers have an edge over their neat counterparts in both domestic and industrial textile applications [161–166].

Polymer clay nanocomposite fibers have been mostly spun through three basic methods of fiber spinning [161]:

(1)  Melt spinning
(2)  Solution spinning
(3)  Electrospinning

The major challenges here are achieving the maximum extent of intercalation and exfoliation of clays in the polymer matrix as discussed earlier (Figure 12). Owing to the hydrophilic character of clay, the preparation of such clay-based nanocomposite fiber is easiest with water-soluble polymers. The increasing nonpolarity and hydrophobic nature of polymers makes it more and more difficult to disperse the hydrophilic clay in the organic matrix. There are two main strategies used to overcome this problem:

(1)  Modification of clay so as to make it organophilic
(2)  Use of a compatibilizer

Organomodified clays have a quaternary ammonium ion with paraffin substituents, for example dimethyl dihydrogenated tallow ammonium chloride (2M2HT) intercalated in interlayer galleries replacing the $Na^+$ and $K^+$ cations making the clays organophilic and thus more dispersible in the polymer system. However, there is a tendency of quaternary ammonium ions to degrade at the higher polymer processing temperatures, that is 150–180°C. Attempts have been made to have groups such as 1,2-dimethyl-hexadecyl-imidazulam-MMT (IMD-MMT) and didecyl-triphenyl-phosphonium-MMT (Cl2 PPH-MMT), which have higher thermal stability [167].

The compounding of MMT clays into a polymeric matrix is usually done on a single or double screw extruder, in which the dispersive forces, that is shearing action inside the barrel, are responsible for breaking the minor components domain, that is clay particle, to the desired size and their homogeneous distribution in the matrix (Figure 16). Most twin-screw processes are fundamentally used for mixing, taking into consideration efficiency, material properties and functionality of the process. For control and optimization of mixing, two fundamental mechanisms are responsible: 'Dispersive Mixing' (reduction in size of the cohesive minor component) and 'Distributive Mixing' (reorganization of the minor component through the polymer matrix), which are determined by the stresses and strains of the polymer flow. The mixing in high viscosity polymers governs two types of mixing flow: shear flow (shear of polymer against a surface) and elongational flow (stretching of polymer melt) (http://www.peptflow.com/default.asp?Lang=1&Page=2). In the case of nanofillers having high aspect ratio such as clay, carbon nanotube or carbon nanofiber, the orientation of nanofillers inside the polymer matrix is of the same importance as its dispersion and distribution in the matrix. It is reported that extensional flow mixer results in better dispersion and distribution mixing and is claimed to be more efficient by orders of magnitude than a shear mixer [161]. Although elongational flow has been demonstrated to be more efficient at mixing, it is more difficult to establish and maintain this flow within a screw mixing device. Hence, the extensional flow mixer has been designed as an inexpensive mixer to be attached to either single or twin-screw extruder.

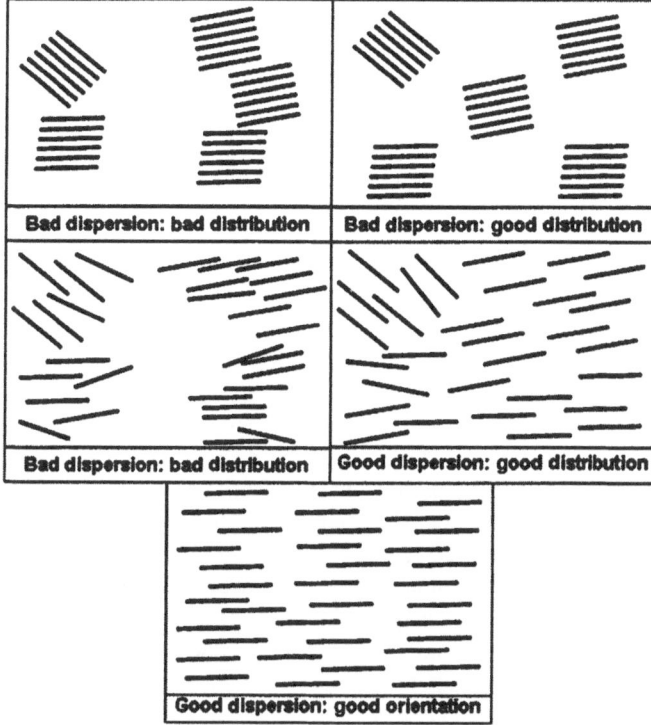

Figure 16.  Schematic representation of distributive and dispersive system of mixing.

All the major classes of polymers used to produce textile fibers, such as nylon, polyester, viscose and PU, have been studied to develop clay-based nanocomposite fibers and their processing and performance are discussed in the following sections.

*4.1.1.1. Nylon/clay nanocomposite fibers.* Polyamide or nylon 6/clay nanocomposites have been widely investigated due to their much superior tensile strength and modulus, improved heat resistance (heat distortion temperature increases from 65°C to 120°C) as well as excellent gas and water barrier properties [168] over their neat nylon 6 counterparts. However, specific studies on spinning and characterization of polyamide/clay nanocomposite fibers have been relatively few [150, 169–178] and are discussed here.

Nylon 6 (PA6)/clay nanocomposite fibers are generally produced using the melt-spinning route, in which nylon 6 chips are compounded with MMT nanoclays during the melt extrusion process using mostly twin-screw extruder. Attempts have also being made to spin fibers of polyamide 6 (PA6)/clay by in-situ polymerization technique using double screw reactive extrusion [150]. The molecular weight and weight distribution measured for the polymerized fibers showed a tendency to first increase and then decrease with increasing screw speed. However, the process succeeds in inserting a covalent linkage between OH groups of organomodified silicate and the amino groups on PA 6.

PA 6/clay nanocomposite fibers have also been produced using the high-speed melt spinning process at take-up velocities of 1–5 km min$^{-1}$. The clay used in all these researches is the most widely used organomodified MMT clay at the 2–5 wt% loading. Careful control of drawing conditions such as drawing temperature and ratio and subsequent annealing step are important to get the desired properties and have been investigated by several researchers [170,172,176,179]. The effect of clay on the fiber structure development and resultant properties has been reported by several researchers [150,170] but in more detail by Giza et al. [176,177] from the Tokyo Institute of Technology, Japan.

PA 6 generally crystallizes in two forms, $\alpha$ form, which is monoclinic, and metastable $\gamma$ form, which is pseudohexagonal. The $\alpha$ form generally grows by slow cooling from the melt or develops on annealing or drawing. The $\gamma$ form is obtained by fast cooling from the melt notably in fibers spun at moderate spinning rates. The nanoclays in PA 6/clay nanocomposite fibers are reported to behave as $\gamma$-nucleating agents in both injection molded samples and oriented spun fibers [178]. However, these rapidly turn into an $\alpha$ crystal form upon drawing. Matrix shearing in between the MMT platelets during drawing imparts better packing and cohesion to hydrogen-bonded planes and thus also improves their thermal stability as compared to those of neat PA 6 fibers [178]. PA 6/clay composite fibers also showed higher crystallinity over the whole take-up range (1–5 km min$^{-1}$) as compared to neat PA 6 fibers (1–7 km min$^{-1}$), although the clay content does not much affect the crystallinity values when produced using high speed spinning. The PA 6/clay fibers when spun at such high speeds show better orientation of crystalline entities and also higher birefringence values, but only in the low take-up region (up to 3 km min$^{-1}$). The orientation-induced crystallization led to direct formation of $\alpha$-form crystals in the spin line, which started at 2 km min$^{-1}$ for PA 6/clay fiber and at 4 km min$^{-1}$ for neat PA 6 fiber as revealed by wide-angle X-ray diffraction (WAXD) studies. The PA 6/clay nanocomposite fibers showed superior Young's modulus at all take-up velocities, but tenacity was higher only in the low-temperature velocity region in which crystallization in the spin line of neat PA 6 did not yet occur [176,177]. Thus, nylon 6/clay nanocomposite fibers show reduced crystallinity and lower molecular orientation, which explains the observation of no significant improvements in tenacity in most of the studies. However, stiffness of intercrystalline regions and the role of rigid clay particles in stress transmission behavior explain the superior modulus in PA 6/clay nanocomposite fibers.

Apart from improved mechanical properties, the presence of clay in PA 6/clay fibers also affects its dyeing ability. The differential scanning calorimetry (DSC) studies show that the presence of clay in PA 6 induces the $\gamma$-crystalline form, increases the crystallization temperature and decreases the melting point. A higher amorphous content observed improves the accessibility for disperse dyes. Thus, PA 6/clay yarn dyes itself faster with disperse dyes as compared with unfilled PA 6 yarn, whereas it gives an opposite effect with acid and metal complex dyes. In both cases, nanoclay fixes on the amino sites, thus preventing the fixation of acid or metal complex dyes [150]. The incorporation of clay into PA 6 fibers also leads to improved fire-resistant and lowered flammability properties, thus offering a new promising route for flame-retardant textiles with a permanent effect and at a low cost [174,175].

*4.1.1.2. Polyethylene terephthalate (PET)/clay nanocomposite fibers.* A number of research studies and patents reported in the literature reveal that there are significant opportunities for enhancing the properties of polyester fibers, mainly polyethylene terephthalate (PET) fibers, by incorporation of inorganic layered silicate nanoclays (MMT) [179–184]. PET/MMT clay nanocomposites are generally prepared by in-situ interlayer polymerization [179–181] and then melt-spun to produce monofilaments with varying draw ratios (DRs). The nanocomposite is prepared by mixing ethylene glycol and dimethyl terephthalate in the presence of organically modified silicate (0.5–15%), 0.001–1.0% additives and/or

0.001–1.0% catalyst (oxide acetate of Sb, Ge, Ti or Sn) and then polymerizing it [180]. The organic modifier used for MMT clays is mostly phosphonium cationic salt [162,179], which makes the organomodified clays thermally stable and thus able to withstand the high temperatures involved during the PET polymerization process. The clay is generally well dispersed in the PET matrix, as observed under scanning electron microscope (SEM) and transmission electron microscope (TEM), but some clays were agglomerated at a size level in the range of 10–20 nm [182]. The interlayer distance of MMT dispersed in the nanocomposite fiber was further increased because of the strong shear stresses during melt spinning [183].

The structure and properties of the PET/clay nanocomposite fibers have been studied as a function of organoclay content and the DR. The PET/clay hybrid nanocomposite fibers generally show improved thermal stability and mechanical properties as compared with their neat counterparts [182] even at low organoclay contents (<5 wt%). The composite fibers also exhibit improved dimensional stability, higher modulus and low shrinkage. The tensile properties generally increase with increasing clay content up to DR = 1. However, the values decrease with increasing DR and also higher clay content beyond a critical concentration [162,184]. DSC and WAXD test results show that the incorporation of clay generally accelerates the crystallization of PET, but the crystallization and orientation of drawn fibers is generally lower than that of drawn neat PET fibers [163]. The strong interaction between MMT and PET restricted the motion of PET chains, which developed a special continuous network structure and inhibited thermal shrinkage of PET and also improved other thermo-mechanical properties of PET fibers.

The PET/clay nanocomposite fibers with improved properties have the potential of finding application in the field of tire cords to reinforce the rubber [164]. High strength, heat resistance, heat-resistant adhesion and low heat emission properties are required for tyres and tyre cords. Polyester tyre cords are more suitable for this use but have some weaknesses. PET/clay nanocomposite fibers can overcome some of these weaknesses of the neat PET fibers and thus enhance the stability, uniformity and ride comfort. The incorporation of clay into polyester fiber also improves its dyeability with disperse dyes, making deep-dyeable polyester a possibility [180].

*4.1.1.3. Viscose/clay nanocomposite fibers.* Unlike PET and nylon 6, viscose is spun through solution spinning method. A project in the University of Dartmouth, USA, headed by Y.K. Kim, has developed a number of nanocomposite fibers and studied their performance with respect to textile areas. It can be concluded from the project that the dispersion of hydrophilic nanoparticles, such as MMT clay, into cellulose solutions is easier than any hydrophobic nanofiller such as carbon nanotube or nanofiber. From a mixing point of view, it is the better approach to produce nanocomposite fibers (www.ntcresearch.org/pdf-rpts/AnRp01/M00-D08-A1.pdf).

Cellulose and regenerated cellulose have the inherent problem of the tendency to catch fire. Several finishes have been developed to improve the flame resistance of cotton and viscose textiles. However, the durability and eco-friendliness of these finishes are a matter of concern. Clay/regenerated cellulose nanocomposites show improved flame retardancy, although the amount of clay loading has to be optimized in order to have textile-grade fibers and filaments [185].

*4.1.1.4. Polyurethane/clay nanocomposite fiber.* The wet spinning of PU fiber is well known to the fiber industry and this fiber has a steady market in the textile industry because of its exceptional properties in terms of stretchability and stretch recovery. However, the strength and dyeability of this fiber are questionable. So, it is generally used as a core of a yarn covered by some common textile fibers such as cotton, viscose or polyester. The problem of PU filaments has been partially overcome using a small amount of clay during its spinning. PU/clay nanocomposite synthesized by solution intercalation has been successfully spun by dry-jet wet spinning method in a doctoral work carried out by our research group. The mechanical properties of nanocomposite filament show optimum values of the tensile and elastic recovery properties at very low clay loadings of 0.25 wt% as compared with neat PU filaments. The use of ultrasonic treatment and high-speed stirring for optimized time results in a combination of intercalated/exfoliated nanocomposite structure. The heat distortion temperature and thermal stability of such nanocomposite filament was higher than that of pristine PU. The most interesting thing is that these nanocomposite filaments have improved dyeability, UV and tackiness resistance [186].

*4.1.1.5. Polypropylene/clay nanocomposite fiber.* The growth of PP as a technical textile material has been quite noteworthy, primarily because of its low cost, inert nature, low density, hydrophobic nature, excellent resistance to chemicals and biological organisms, and a wide range of physical properties. However, some of its shortcomings are a low use temperature, difficulty with dyeing, poor UV resistance and a tendency to creep at and above room temperature. The recent tremendous success of polymer/clay hybrid nanocomposites offers attractive potential for the continuous expansion of the application versatility of PP for industrial textile use.

PP/nanoclay composite filaments have been produced by melt spinning and the effect of the compatibilizer and the role of the nanoclay are studied thoroughly for improving the physical properties. The use of maleic anhydride grafted PP compatibilizer eases the mixing and dispersion of nanoclay to a large extent. Clay loadings up to 1 wt% with up to 3 wt% compatibilizer were used and a significant improvement in the tensile, thermal and dynamic mechanical properties and creep resistance of PP/nanoclay composite filaments was achieved over neat PP filaments at very low clay loadings of 0.25–0.5 wt% and a compatibilizer/clay ratio of 2:1 in one of the studies carried out by our research group. It has been demonstrated that compatibilized PP/clay nanocomposite filaments can be successfully spun and drawn into a filament form up to 1 wt% clay loading. The fibers can be drawn to a very high DR with a three-stage drawing process. The incorporation of the nanoclay into the PP matrix, especially in the presence of a compatibilizer, leads to a tremendous improvement over a wide range of properties at very low weight fractions of the clay. A marked improvement in the tensile properties, that is, the tensile strength (30%) and tensile modulus (70%), and in the dynamic mechanical properties, that is, the storage modulus, tan $\delta$ and creep resistance, and reduced thermal and boiling water shrinkage have been achieved at very low loadings of the clay (0.25–0.5 wt%) and at a compatibilizer/clay ratio of 2:1 in comparison with neat PP fibers in a research study carried out by M. Joshi and V. Vishwanathan [187]. These superior performance properties of the PP/nanoclay compatibilized filaments provide an edge over conventional PP for use in various technical textile applications.

Modification of pure sodium MMT nanoclay can be carried out using any long chain paraffin substituents such as hexadecyl trimethyl ammonium bromide (HTAB) via an ion exchange reaction. The extent of modification can be characterized through X-ray diffraction, as the modification of MMT increases the interlayer d-spacing of clay gallery from 1.22 to 1.83 nm. The modified clay can be melt blended with PP in the presence of a swelling agent. The spinning and drawing conditions of such PP/clay nanocomposite filaments were strongly influenced by the different weight percentages of nanoclay. It has been observed that the nanoclay reinforced PP nanocomposites can be spun and drawn successfully for 0.5, 1.0 and 1.5 wt% of the modified clay loading. Beyond 1.5 wt%, the spinnability is poor. The composite PP filaments with modified clay show improved tensile strength, modulus and reduced elongation at break. The 0.5 and 1.0 wt% reinforced PP/modified nanoclay composite filaments show a significant increase in tenacity (∼30%) as well as modulus (∼100%) as compared with neat PP filaments. The modified clay content in the range of 0.5–1.0 wt% gives optimum properties. Any improvement in tensile strength cannot be observed in the case of composite filaments with unmodified clay, but the modulus improved as the clay matrix interaction is less, while the stiffness of the filaments increased. The improved creep resistance and thermal stability was observed in filaments with modified as well as unmodified clays.

All these properties can be correlated to the morphology of the nanocomposite filaments. The X-ray diffraction of the samples showed enhanced crystallinity in the presence of modified clay. The sharp and narrow X-ray diffraction peaks of PP/nanoclay composite filaments indicate an increase in crystallinity in the presence of modified clay at small loadings (0.5%). The scanning and TEM studies indicate a good dispersion of modified clay in the PP. This has facilitated the homogeneous dispersion of clay in the PP matrix forming a combination of intercalated and exfoliated structure, responsible for enhanced mechanical and thermal properties [188]. However, true molecular dispersion resulting in a completely intercalated/exfoliated composite structure is expected through the use of a twin-screw extruder for intensive melt mixing and suitable compatibilizers.

### 4.1.2. Polymer/carbon nanotube nanocomposite fibers

Because of the immense potential to impart unique mechanical, electrical and thermal properties to polymeric fibers, polymer nanocomposites dispersed with CNTs and nanofibers are beginning to receive substantial attention in the scientific community [189].

Both single-wall CNTs (SWNTs) and multi-wall CNTs (MWNTs) are characterized by high flexibility, low mass density and large aspect ratio, that is 300–1000. Some nanotubes are stronger than steel, lighter than aluminum and more conductive than copper. The theoretical and experimental results on SWNTs also show extremely high tensile modulus (640 GPa–1 TPa), tensile strength (150–180 GPa) and failure strains, of the order of 5–20%. SWNTs also possess very high electrical and thermal conductivity (theoretically >6000 W/mK). Depending on their structural parameters, SWNTs can be metallic or semi-conducting [189]. The first nanocomposite based on CNTs was reported by Ajayan et al. in 1994 [190,191]. The major challenges in polymer/CNT nanocomposites are:

(1) 'Less than ideal' and inconsistent nature of CNTs. All known methods of CNT production give a mixture of CNTs with different chiralities, diameters and lengths, along with varying amount of impurities and structural defects within the sample or that from different batches. Hence, it is very difficult to obtain reproducible results [189].
(2) Dispersion of CNTs in the polymer matrix is also very crucial for achieving the predicted properties. Due to van der Waals interaction, SWNTs typically form bundles containing several hundred tubes within a bundle. These tubes are also highly entangled and the degree of entanglement depends on their length. Various physical and chemical approaches have been tried to disperse them, which include ultrasonication, choosing the correct dispersion medium and use of surfactants to aid in dispersion [192].

(3) Another great challenge is the efficient translation of CNT properties into the polymer matrix, which means ensuring a good interfacial interaction between the two. CNTs being inert are not easily dispersed or interact with most of the polymer matrix systems.

The approaches used are to functionalize the CNTs by oxidation, end cap functionalization with long aliphatic amines and sidewall functionalization using fluorination, alkylation, and so on [193]. Polymer grafting to create functional groups on CNTs has also been reported, which improves their dispersion as well as interfacial interaction. Noncovalent functionalization achieved by adsorbing different polymers onto SWNT to improve the SWNT solubilization by 'polymer wrapping' is also an alternative method for tuning the interfacial properties of nanotubes [194].

Carbon nanotube orientation preferentially along the fiber axis plays a very critical role in imparting high tensile properties in CNT-based polymeric nanocomposite fibers. SWNTs with a density of 1.3 g/cm$^3$ have the potential of producing a high-performance fiber with at least two times the specific tensile strength and specific tensile modulus of the presently known state-of-the-art polyacrylonitrile (PAN)-based carbon fiber [189,192,195] even at a very low loading. However, large-scale economic synthesis of CNTs and the ability to synthesize CNTs of uniform diameter and chirality will be critical to the commercial success of these new advanced classes of materials for moulded parts, films as well as fiber forms.

As one of the obstacles to the widespread use of application of the nanotubes is the inability to orient them in one particular direction, oriented polymer systems such as fibers provide a unique opportunity to align the CNTs in one particular direction during the process of melt spinning and subsequent drawing step. Different polymer/CNT-based nanocomposite fibers are discussed below.

*4.1.2.1. Nylon/CNT-based nanocomposite fibers.* There are several reports on nylon/CNT nanocomposites, in which CNTs (single or multiwalled) or CNFs (carbon nanofibers) are infused into the polymer through a liquid route using ultrasonication or a dry route followed by melt mixing in a single or twin-screw extruder [196]. Alignment of CNTs and CNFs in the compositions was enforced during extrusion or subsequent stretching process. However, the fullest potential of CNT reinforcement has not been harnessed in polyamide/CNT nanocomposite fibers, primarily because of the lack of alignment or failure to develop strong interfacial bonding between CNTs and polymer.

Meng and coworkers [197] report a method to fabricate nylon 6/CNT reinforcement, in which the acid and diamine-modified MWNT reinforcement at low loading (<1 wt%) leads to an increase in the modulus. The enhancement of mechanical properties is attributed to the high interfacial shear strength developed during the fabrication process. These polyamide/CNT nanocomposite fibers have a great potential for use as multifunctional textile materials in several technical and advanced composites applications [197].

*4.1.2.2. Liquid crystal polymer/CNT-based nanocomposite fibers.* Jose and coworkers at the University of Alabama at Birmingham [198] focused on aligning CNTs in Vectra (a thermotropic liquid crystalline polymer) as well as PP matrix during melt spinning into fibers. CNTs in two different weight percentages (0.5% and 1.0%) were used. Correlation of CNT alignment along the fiber axis and dispersion in the polymer matrix was studied using TEM. A significant improvement in mechanical properties, that is tensile strength and modulus, and a substantial increase in the onset of decomposition temperature indicated both a good dispersion and a highly aligned CNT system.

An inherent hierarchical structure was observed in the form of fibrils that develop in vapor-grown carbon nanofibres (VGCFs) dispersed in a liquid crystal polymer matrix by Rohatgi et al. [199]. The tensile properties were sensitive to the diameter of extruded filament, the amount of VGCF added and other parameters related to the extrusion process. There is a noticeable increase in mechanical properties of composite filaments with decreasing filament diameter irrespective of VGCF content.

*4.1.2.3. Polyurethane/CNT-based nanocomposite fibers.* Nanocomposite fibers based on thermoplastic PU and MWNTs show a significant improvement in Young's modulus and tensile strength achieved by incorporating MWNTs up to 9.3 wt% without sacrificing the PU elastomer's high elongation at break. The nanotubes were functionalized for a better interaction with the polymer matrix before the dispersion. A twin-screw extruder has been used to disperse MWNTs into the polymer and then to extrude the nanocomposite fiber. Electron microscopy results indicated that the homogeneous dispersion of MWNTs throughout the PU matrix and the strong interfacial adhesion between oxidized MWNTs and the matrix are responsible for the considerable enhancement of mechanical properties of the composite fibers [200].

*4.1.2.4. Polyethylene/CNT-based nanocomposite fibers.* Controlled polymer crystallization confined in nanotube aerogel fibers leads to an exceptional increase in tensile and electrical properties. The investigation on polyethylene/CNT nanocomposites by means of polarized optical microscopy (POM), SEM and WAXD reveals that the individual nanotubes are periodically decorated with polyethylene nanocrystals, forming aligned hybrid shish-kebab nanostructures. After melting and recrystallization, transcrystalline lamellae connecting the adjacent aligned nanotubes have been developed. Microstructural analysis shows that the nanotubes can nucleate the growth of both orthorhombic and monoclinic crystals of polyethylene in the quiescent state. The tensile strength, modulus and axial electrical conductivity of such polyethylene/CNT composite fibers are as high as 600 MPa, 60 GPa, and 5000 S/m, respectively [201].

*4.1.3. Other nanoparticle-based nanocomposite fibers*

Preparation of PA6/nano-$TiO_2$ composites and their spinnability have been reported by Zhu et al. [202] of Dong Hua University, Shanghai, China. The nanoscaled $TiO_2$ particles were surface treated with coupling agents prior to mixing with molten PA 6. The composite fibers showed improved mechanical properties as compared with pure PA 6 fibers and composite fibers with unmodified $TiO_2$.

PET/silican dioxide ($SiO_2$) nanocomposites were synthesized using in-situ polymerization and melt spun into fibers. The PET/$SiO_2$ nanocomposite [203,204] fibers showed a greater degree of weight loss as compared with pure PET fibers when they were hydrolyzed using alkali treatment. Superfine structures such as cracks, craters and cavities were introduced due to this facilitated deep dyeing of PET/$SiO_2$ nanocomposite fibers. In spite of these defects, PET/$SiO_2$, as spun nanocomposite fibers, shows improved tenacity, modulus and lower heat shrinkage [205]. The silica nanoparticles get well dispersed in PET in the range of 10 nm with a narrow distribution.

PET/$TiO_2$ nanocomposites fibers have been prepared by in-situ polycondensation and melt spinning [206], Nanotitania were first treated with a coupling agent to introduce functional groups on the surface of titania particles and thus aid in the homogeneous

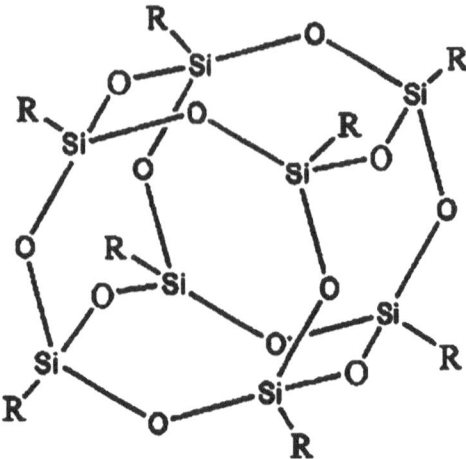

Figure 17. Molecular structure of POSS (R- any organic group) [208]. Reprinted from M. Joshi and B.S. Butola, J. Appl. Polym. Sci. 105 (2007) pp. 978–985, with permission from Wiley Inter Science, John Wiley and Sons.

dispersion of nanoparticles. The UV production property of these fibers was much enhanced (UPF >50) because of high refractive index and absorption of UV light by rutile form of nanotitania incorporated. However, these fibers had a slightly reduced tensile strength and elongation at break. PET nanocomposite fibers have also been reported with a range of other nanoparticles, that is $TiO_2$, $SiO_2$, ZnO and $CaCO_3$. The nanoparticles are treated with a low surface tension additive before mixing with PET [207].

POSS are an important class of nanostructured hybrid inorganic-organic materials and are widely considered as some of the most promising and rapidly emerging nanocomposite materials. POSS molecules have a nanosized polyhedral or a cage-type structure consisting of an inner silicon–oxygen $(SiO_{1.5})_x$ based framework, with organic substituents (R) on outer corners and with a precisely defined silicon–silicon distance of 0.53 nm (Figure 17).

The ratio of Si:O is 2:3 and the average diameter of a POSS molecule is 1.5 nm. The inorganic silicon–oxygen core is thermally and chemically robust. The organic substituents (R) are either totally hydrocarbon in nature or a range of polar structures and functional groups are also included. These organic substituents can be totally reactive or nonreactive or a combination of the two, depending on the synthetic design. According to the nature of the organic substituents, they undergo nucleophilic substitution or hydrosilylation. The density of POSS chemicals is typically in the range of 1.12 gm/ml, despite a high molar mass of about 1000 amu. The reason is that most of the volume is taken up by the organic (R) groups, which lie on the outside of the cage, whereas the core occupies only 5% of the overall cage volume. The physical state (solid, wax or liquid) ranges from a low solubility system with a $T_m$ of approximately 400°C to wax or grease or low-temperature flowable oil depending on the nature of the organic (R) groups along with the topology of POSS cages [208].

POSS were first developed by the U.S. Air Force for aerospace applications. The original cost of POSS was $5000 per pound and took up to three years to produce. Due to simplified and redesigned process chemistry, the production cost has now come down, but still it is expensive. A variety of POSS monomers in the form of solids or oils are now commercially available from Hybrid Plastics Company Co. (http://www.hybridplastics.com/), USA. A wide range of POSS compounds are now available that contain a combination of covalently bonded reactive functional groups that make them suitable for polymerization, surface bonding or other chemical transformations [209]. POSS moieties can thus be easily incorporated via copolymerization, grafting or blending into common plastics. POSS-incorporated polymeric materials show significant improvements in use temperatures, oxidation resistance, surface hardening and enhanced mechanical properties. Reductions in flammability, heat evolution and viscosity during processing are also the additional advantages [210].

POSS-based nanofillers have been incorporated in PET and polyamide fibers/filaments, and show improved thermo-mechanical properties and better retention of modulus at higher temperatures. These fibers can find an application in automotive tyre reinforcement [211]. However, the extent of reinforcement was found to be variable in the case of PET systems, which can be either due to a low interaction between PET and POSS causing processing problems related to some water generation or due to chemical transformation of POSS particles themselves [212].

The conducting and semi-conducting flexible and stretchable filaments are in a great demand for biomedical and defense applications. The metal threads lack stretchability while the conducting polymers do not form fibers. Before the advancement of nanocomposite fibers, the only alternative was the coating on fibers, which is not at all durable and robust and lacks the capacity of generating steady pulses over a long time. Here comes the role of nanocomposite fibers having conducting nanoparticles dispersed into it. Conducting nanofillers such as carbon nanotube/nanofiber and nanographite have the capacity to improve the conductivity of polymers without affecting the properties of the bulk polymers. After a rigorous mixing of a very small amount of such nanofillers in the twin screw, the conductivity of PU, which is as such nonconducting, can be enhanced by more than $10^4$ times [213].

## 4.2. Nanofibers

Nanofiber is a new class of nanomaterials having a diameter in the submicron range (50–500 nm), whereas length is in the micron or millimeter range. Due to their very high surface area, they possess some unique properties that make them potential candidates for a wide range of applications such as filtration, barrier fabrics, protective clothing, wipes and biomedical applications such as scaffolds for tissue engineering [214]. Some critical applications such as chemical or biological protective suits need the electro-deposition of nanofibrous material on their seams to prevent the attack through these weakest areas of the garment. Although nanofibers can also be spun through other ways, for example rotary jet spinning and vapor growing (CNFs and nanotubes), electrospinning is the most popular method of nanofiber spinning.

### 4.2.1. Electrospinning

The manufacturing technique most commonly associated with polymeric nanofibers is electrospinning, which is fundamentally different from conventional fiber production techniques and is based on electrostatic forces. Conventional fiber spinning techniques, for example melt spinning, dry spinning or wet spinning, rely on mechanical forces to produce fibers by extruding polymer melt or solution through a spinneret and subsequently drawing the resulting filaments as they solidify or coagulate. By introducing a high electric field to modify the fiber formation process, it is found that it is possible to develop smaller-than-conventional diameter fibers with a better uniformity throughout the electrospinning technique.

The effect of electric field on liquid droplet was studied long ago by Rayleigh (1882) and Zeleny (1915). Since then, a number of researchers concentrated their experiments to produce a fiber from a polymer melt or solution with the help of an electric field. The principle of electrospinning is to use an electric field to draw a positively charged polymer solution from an orifice to a collector. This creates a jet of solution from the orifice to the grounded collection device. The jet emerges at the base from the nozzle, which has the geometry of a cone (Taylor cone). Then, it travels to form a stretched jet and divides into many fibers in the splaying region. But splaying may not always be the case. A rapidly rotating spiral jet in a whipping motion, which is indistinguishable from the splaying phenomenon to the naked eye, was also observed [215]. Whatever is the mechanism, the fibers are eventually collected on a grounded metal screen and the diameters of these fibers are typically hundreds of nanometers, one to two orders of magnitude smaller than the fibers produced by conventional extrusion techniques [216].

*4.2.1.1. Background.* The patent by A. Formhals in 1932 on electrostatic spinning of acetate rayon fibers describes the formation of elongated drop under high electric field resulting into very fine fibers or filament [217]. An electrode connected to a high voltage power supply is inserted into a polymeric solution contained within a capillary tube. A grounded collection plate or screen is placed at a distance from the tip of the capillary tube. Initially, the polymer solution is held by its surface tension in the form of a droplet at the end of the capillary tube. As the voltage is increased, charge is induced on the fluid surface, and the droplet is distorted to a conical shape (not the spherical shape as in conventional melt spinning). Above a critical voltage, a single jet is ejected from the apex of a conical meniscus. Later, this conical meniscus came to be known as the 'Taylor cone'. Beyond this conical base immediately at the end of the capillary tube, the jet continues to thin. This jetting mode is known as the electrohydrodynamic cone-jet.

Although the idea dates back to 1932, recent interest has been renewed by the pioneering work done by Dr. Darryl Renekar at the University of Akron, USA [218], who has demonstrated electrospinning for a wide variety of polymer solutions including rigid rod polymers. The basic theory of the nanofiber spinning process and the parameters affecting the process, thermal and mechanical properties of electrospun nanofibers have also been studied extensively [219,220].

*4.2.1.2. Process.* Electrospinning is a process that produces continuous polymeric nanofibers (diameter in the submicron range) through an action of an external electric field imposed on a polymer solution or melt. In this technique, a polymer is dissolved in a solvent (polymer melts can also be used) and generally placed in a glass capillary, which is sealed at one end and a small opening in a necked-down portion at the other end. A high voltage potential (5–50 KV) is then applied between the polymer solution and a collector or target at the near end of the capillary tube (Figure 18). When a high voltage is applied to the polymer solution, the electric force results in the formation of a jet of polymer solution flowing out from a droplet tip to be drawn toward a grounded collector.

After the jet flows away from the droplet in a nearly straight line, it bends and whips into a complex path and other changes in shapes occur, during which high electrical forces stretch and thin it at very large ratios (about 100,000) in a short distance and in less than 1 s. After the solvent evaporates, solidified nanofibers are left, mostly in the form of a nonwoven web on the collector. This nanofiber deposition cannot be seen by the naked eye and is generally viewed under the SEM (Figure 19).

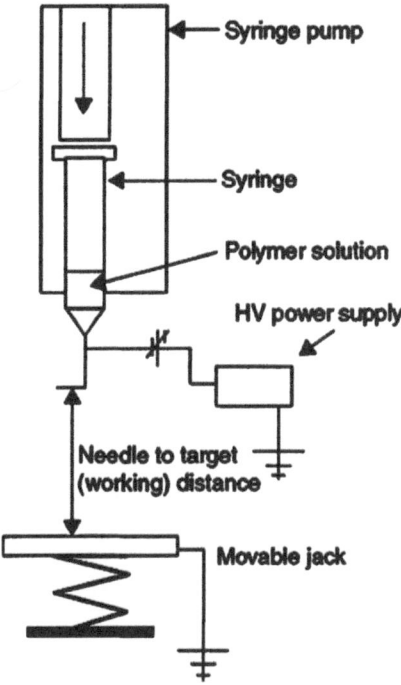

Figure 18. Schematic diagram of electrospinning set up [221]. Reprinted with permission from S. Megelski, J.S. Stephans, D.B. Chase and J.F. Rabolt, Macromolecules 35 (2002) pp. 8456–8466. Copyright 2002, American Chemical Society.

The process can be considered as a variation of the better known electrospraying process. In the electrospraying process, the surface of a hemispherical liquid drop suspended in equilibrium at the end of the capillary is distorted into a conical shape in the presence of an electrostatic field. The distortion is due to the balancing or responsive forces resulting from the induced charge distribution on the surface of the drop with the surface tension of

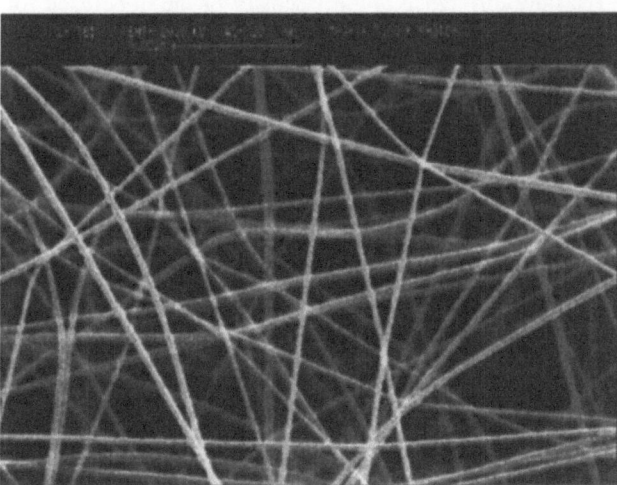

Figure 19. Electrospun nanofibrous web under SEM.

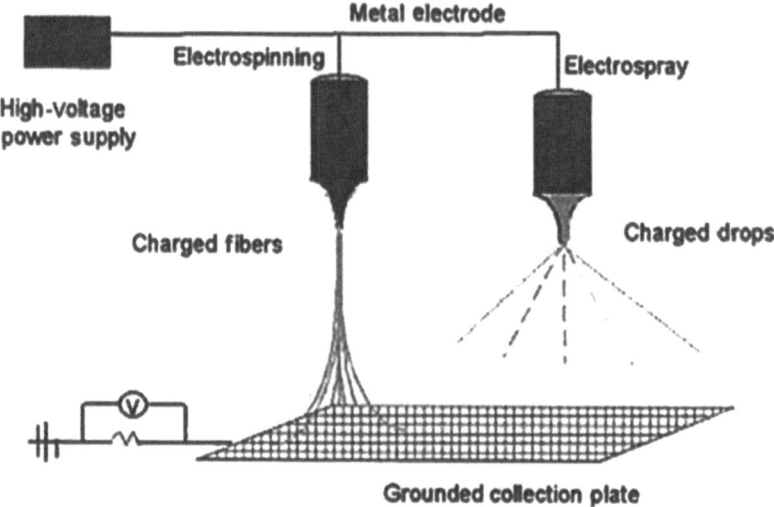

Figure 20. Electrospraying and electrospinning set up [222]. Reprinted from M. Deitzel, J. Kleinmayer, D. Harris and N.C. Beck Tan, Polymer 42 (2001) pp. 261–272, with permission from Elsevier.

the liquid. Once a critical voltage, $V_c$, is exceeded, a stable jet of liquid is ejected from the cone tip. The jet breaks into droplets as a result of low surface tension in the case of low viscosity liquids. However, for high viscosity liquids, the jet does not break up but travels to the grounded target and thus produces ultra-thin fibers by constant stretching until it is grounded to dissipate the charge. The first case is known as electrospraying and is used in many industries to obtain aerosols composed of submicron drops with narrow distributions. When applied to high-molecular weight polymer solutions and melts, it generates continuous polymer fibers that are submicron in diameter, and this second case is known as electrospinning [222]. Figure 20 describes the similarities between the two processes. The only difference is the viscosity of the dope used for the purpose. If the viscosity and surface tension of the dope is suitable to withstand the tremendous force exerted by the high voltage, it deposits as a fiber otherwise minute droplets are formed.

The idea to spin textile-grade fiber using this principle could not be developed for a long time, due to lack of the theoretical foundation, poor understanding of the process and consequent limitation in process control, reproducibility and productivity. In 1969, Taylor derived an equation for predicting the critical voltage for monomeric liquids at which the fine jet would first appear. Experiments have been carried out to find whether this principle would be applicable to produce continuous filaments from high-molecular weight polymers.

Most of the literature on electrospinning has explored a variety of polymer/solvent systems from which fibers can be produced. Only a few studies have addressed the processing/property relationships in electrospun fibers. The structure and property of electrospun nanofibers are predominantly determined by the synergistic effect of solution parameters and electrostatic forces. Solution concentration and viscosity effects, spinning atmosphere effects, accelerating voltage effects and tip-to-target distance are the studied processing parameters [222,223]. Solution viscosity has been found to influence fiber diameter, initiating droplet size and jet trajectory. Increasing solution viscosity has been associated with the production of large fiber diameters. Spinning atmosphere has been associated with the jet splaying phenomenon, also reported by Srinivasan and Renekar [224]. Other processing variables, such as acceleration voltage, electrospinning current and tip-to-target distance, have also been investigated and linked to fiber morphology and defect structures [225].

*4.2.1.3. Taylor cone formation.* When the polymer leaves the orifice, it forms a cone under the high electric field before thinning to produce a jet; this cone is named as the 'Taylor cone' after the name of the inventor. Studies with polyethylene, PP in melt and polyethylene in paraffin solution show that the meniscus formed just before the jet formation is nearly conical with a semi-vertical angle of approximately 50°. The change of shape of the Taylor cone from convex to concave is in accordance with changes in the charge density as the field strength is varied. However, the thinning jet is relatively insensitive to the details of the Taylor cone [226]. Analysis of the flow field in an electrically driven jet revealed that the flow is a combination of parabolic and purely extensional flow. The extensional strain rate along the symmetry axis was found to depend on the electric field intensity at which the jet was operating. So, continuous oriented fiber spinning is possible by appropriately chosen parameters [227].

Earlier, it has been found that the hyperboloidal approximation employed permits prediction of the stationary critical shapes of drops of inviscid, Newtonian, viscoelastic and purely elastic fluids. It was shown, both theoretically and experimentally, that, as a liquid surface develops a critical shape, its configuration approaches the shape of a cone with a half angle of 33.5°, rather than a Taylor cone of 49.3°. The critical half angle does not depend on fluid properties for Newtonian fluids, as an increase in surface tension is always accompanied by an increase in the critical electric field. However, the sharpness of the critical hyperboloid depends on elastic forces and surface tension in elastic fluids as well as in unrelaxed viscoelastic fluids [228].

*4.2.1.4. Approaches to produce a nanofiber.* A typical jet flow is shown in Figure 21 under the electrostatic field; the polymer drop first forms a conical shape that continues

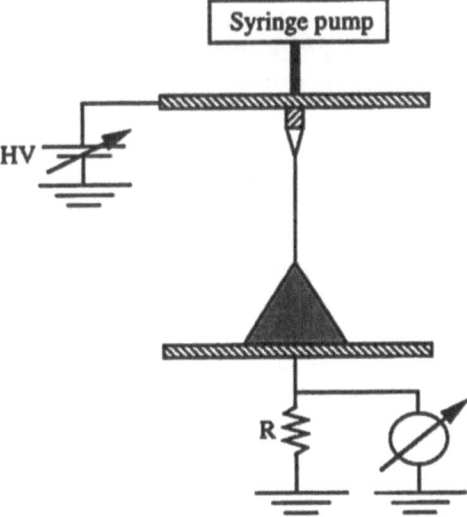

Figure 21. A typical jet in electric field [216]. Reprinted from Y.M. Shin, M.M. Hohman, M.P. Brenner and G.C. Rutledge, Polymer 42 (2001) pp. 9955–9967, with permission from Elsevier.

to a stable filament formation to some distance, with the process of continuous thinning the same as the elongational flow deformation under the take-up force in conventional spinning. Jet profile is certainly not the pressure-driven flow but the electrically driven flow. Then, after covering a certain distance, radial charge repulsion results in splitting of the primary jet into multiple filaments, in a process known as 'splaying'. In this view, the final fiber size is determined primarily by the number of subsidiary jets formed. High pressure application during melt spinning often leads to flow instability especially in cases where high orientation is required. To produce small denier fibers from high-molecular weight materials or with materials having strong polar bonding, conventional elongational flow field is unsuitable. To overcome the limitation of inconsistent strain field generation in spinning of small diameter fibers, attempts were made to elongate them under electrostatic force. Melt coming out from the deformed droplet as a jet was threaded through an aperture on the metallic plate and wound to a spool rotating at an appropriate speed [226]. The plate and capillary were placed at a distance of 1–3 cm and the applied electric field was 3–8 kV cm$^{-1}$. The stable jet region is very small, so the rate of crystallization of the polymers used for this type of filament production must be very high. That is why initial experiments are carried out with polyolefin (both melt and solution) and acrylic (in solution). The following observations on the diameter of filament were made:

- Width of fibers decreased with an increase in applied voltage.
- The diameter of fibers decreased with the increase in melt temperature.
- The higher the take-up speed, the lower the diameter.
- With the increase in separation distance from the orifice, the diameter shows a rapid decrease and then it becomes constant. The initial decrease is the combined effect of the forces due to gravity and electrostatic stress. When these forces are balanced by viscous effects and aerodynamic tangential drag, the diameter stops decreasing.

The estimate on spinning rate of the experimented polyolefin was found in the order of 1 m min$^{-1}$. For a particular temperature (200°C), it was found that the relationship between the spinning rate and electric field strength $E$ is of the form, rate $= Ae^{bE}$, where A and b are constants with values of 38.7 and 0.16, respectively, (for polyethylene melt), if the rate is expressed in cm min$^{-1}$ and $E$ in kV cm$^{-1}$.

In conventional high-pressure fiber extrusion process, the spinning rates are of the order of 1000 m min$^{-1}$. From the above equation, the speed can be achieved at about 50 kV cm$^{-1}$, but in practice, the maximum electrostatic spinning rates are limited by several other factors such as shortening of the stable jet region and breakdown of the air in the gap between the electrodes [226]. Further, experiments carried out on polyethylene and nylon 12 show that the deformation behavior of polymer melt under a high electric field follows the theory developed for Newtonian fluids. This is surprising considering the non-Newtonian behavior of the polymer melts. It has also been proved theoretically that all the fiber-forming polymer melts can be used to produce electrospun nanofibers [229].

The effect of surrounding gas during acrylic spinning was found to be interesting. The air with different humidity, helium and Freon-12 values was used to study the effect. In dry air, the spinning drop tended to dry out and spinning could be carried out for only 1–2 min. In high humid air (>60%), the fiber did not dry properly and tangled above the grounded screen. No spinning could be done in a helium atmosphere, as the gas started breaking down electrically at 2500 V. Fibers spun in Freon-12 gas were from 1.4 to 2.6 times the diameter of fibers spun in air at identical conditions. Small fiber offshoots originated in the first few centimeters of the main fiber. They all seemed to originate at, or be accompanied by, sharp bends in the main fiber. It was assumed that this is due to the higher breakdown voltage of Freon-12 as compared with air [230], a phenomenon also observed by other authors [225]. However, the equation for prediction of critical voltage is as given below [226]:

$$V_c^2 = A\,H^2/L^2(\ln\,2L/R - 1.5)(0.117\Pi R\gamma)$$

where $H$ = distance between the electrodes, $L$ = length of the capillary, $R$ = radius and $\gamma$ = surface tension.

*4.2.1.5. Instability of jets.* To understand the electrospinning process, the most crucial element is the competition between the modes of instability. Whether whipping or axisymmetric breakup of the jet dominates for a given set of conditions depends on the jet radius and the surface charge experienced by a fluid element as it travels downstream. The rapid growth of a nonaxisymmetric, or 'whipping', instability causes bending and stretching of the jet during electrospinning [215].

When a viscoelastic liquid drop is subjected to a high electric field, the simple idea is that the jet lengthens in a straight line along its axis at an incredibly high velocity until its leading end touches the target or collector and dissipates all charges before coagulating as a fiber. Obviously, if this is the case, the resulting fibers' diameter would be at a minimum in the micron range. However, instead of directly coming onto the collector plate, the jet bends and develops a series of lateral excursions that grow into spiraling loops. Each of these loops grows larger in diameter as the jet grows longer and becomes thinner. About 20 ms after a loop has formed, a new set of electrically driven bending instabilities appears on the now thinner, smoothly curved loop, occurring within a time interval of less than a millisecond. The new bending instabilities grow when the jet in the loop is thin enough and when the viscoelastic stress along its axis has relaxed enough. These new bending instabilities develop into a smaller set of spirals that loop around the path of the first loop. The envelope cone observed at the end of the straight segment defined the region inside which this complex path of the jet has developed. The cycles of bending instability repeat in a self-similar manner until the solvent is evaporated, and the remaining polymer fibers resist further elongation by the Coulomb forces of the charge that were still present on the jet. Finally, it is deposited on the collector, which is grounded to dissipate all the residual charges present in the fiber.

The simple case of a single, uniformly thinning jet extending from the nozzle to the collector has been observed at low electric fields (Figure 22a). At high electric fields, after traveling a short distance, the jet becomes unstable. With regard to the exposure times down to 1 ms, the unstable region of the jet has the appearance of an 'inverted cone', suggestive of the envelope created by multiple jets (Figure 22b). Using high-speed photography, it has been confirmed that the inverted cone image is in reality a single, rapidly whipping jet (Figure 22c). The whipping frequency is so fast that the jet appears to be splitting into multiple filaments, a phenomenon observed earlier for electrostatic sprays. The production of submicron-diameter fibers is always preceded by the onset of this whipping instability. The whipping instability is one of the several possible instabilities that may occur in an electrified fluid jet. Additional types of instabilities may lead, for example, to the breakup of the jet into droplets. Interactions between the free charge in the jet and the external electric field give rise to a competition between the different instabilities, which are convected downstream and grow at different rates, depending upon the fluid parameters and the operating conditions. Controlling electrospinning therefore requires a model by which the

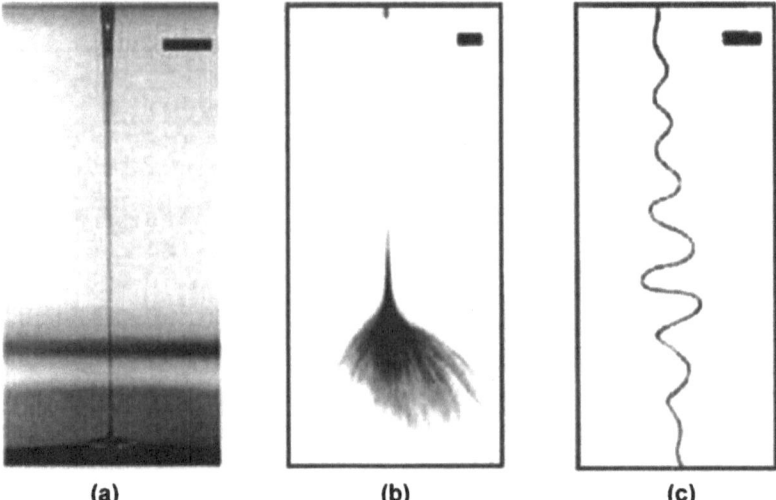

**(a)**               **(b)**               **(c)**

Figure 22. Jet images of a 2 wt% solution of PEO (MW = 2000000) in water during electrospinning. (a) Stable jet ($E_\infty = 0.5$ kV/cm, $Q = 4$ ml min$^{-1}$, scale bar = 5 mm); (b) unstable jet ($E_\infty = 0.67$ kV/cm, $Q = 0.1$ ml min$^{-1}$, scale bar = 5 mm, 1 ms exposure); (c) close-up of the onset of instability ($E_\infty = 0.67$ kV/cm, $Q = 0.1$ ml min$^{-1}$, scale bar = 1 mm, 18 ns exposure) [215]. Reprinted with permission from Y.M. Shin, M.M. Hohman, M.P. Brenner and G.C. Rutledge, Appl. Phys. letters 78 (2001) pp. 1149–1151. Copyright 2001, American Institute of Physics.

competition between the fluid instabilities in an electrified fluid jet can be quantified. From this, operating conditions can be found that take advantage of the desired instability.

Looking into the inside of such bending instability, it has been found that the initial straight segment of a typical electrically charged jet was created by the electrical potential applied between the pendent drop and the collector. The longitudinal stress caused by the external electric field acting on the charge carried by the jet stabilized the straight jet for some distance. Then, the lateral disturbance grew in response to the repulsive forces between adjacent elements of charge carried by the jet. The motion of segments of the jet grew rapidly into an electrically driven bending instability. Then, an electrically driven bending instability, triggered by perturbations of the lateral position and lateral velocity of the jet, grew. The repulsive forces between the charges carried with the jet caused every segment of the jet to lengthen continuously along a changing path until the jet solidified. A very high reduction, as much as $10^5$ times in the cross-sectional area of the jet, can be achieved in this manner and the corresponding large increase in the jet length occurred in a region that was only a few centimeters across. The associated high longitudinal strain rate implies that, in the electrospun jet, the macromolecular coils are stretched along the axis of the jet.

Looking into the dynamics of thin free liquid jets moving in air with a high speed, the difference in the aerodynamically driven and electrically driven instabilities is essentially the destabilizing force. The bending force acting on the jet is due to the mutual Coulomb interactions. It has been proven that the bending instability of jets in electrospinning is responsible for the formation of nanofibers and a very general basic instability of a system of charged particles corresponds to the Earnshaw theorem of electrostatics. This theorem leads to the conclusion that it is impossible to create a stable structure in which the elements of the structure interact only by Coulomb's law. Charges on or embedded in a polymer fluid move the fluid in quite complicated ways to reduce their Coulomb interaction energy. Electrospinning utilizes this behavior to produce interesting and useful polymer objects [231]. However, the solvent evaporation rate and solidification of the polymer jets are two important parameters leading to the formation of dry nanofibers. Thus, the nature of solvent is also a factor to be studied in detail. Considering the evaporation and solidification, the shape of the envelope cone that surrounds the bending loops of the jet in the course of electrospinning can be figured out. Also, the downward velocity of the jet can be calculated to be within an order of magnitude of the observed velocity. The theoretical results also allow for the calculation of the elongation of material elements of the jet. The calculated results also illustrate the increase in viscosity of segments of the jet as the solvent evaporates during the course of electrospinning. It is emphasized that currently, information on the rheological behavior of polymer solution being elongated at the rate and other conditions encountered during electrospinning are rather scarce. Data on evaporation and solidification of polymer solutions in the electrospinning process are practically unavailable. Therefore, at present, a number of the parameters in this calculation can only be estimated by the order of magnitude or can be found from experimental observations of the electrospinning process. Material science data acquired for the electrospinning process will allow researchers to avoid such obstacles in the future [232].

The stability characteristic of the jet is a function of realistic and measurable values of the fluid properties and operating parameters. The most important characteristics of the fluid are found to be viscosity and conductivity. The main operating parameters are the applied electric field and flow rate. The relevant process variables are the jet radius, axial velocity, charge distribution and displacement of the centerline of the jet as a function of the axial distance from the nozzle. In general, the whipping mode dominates at a high charge density, whereas the axisymmetric mode dominates at a low charge density. The instability equations cover a wider range of materials and process parameters, and are expressed in terms of well-known fluid properties [215].

*4.2.1.6. Collectors.* The shape of the collector is most important in terms of the orientation of nanofibers in the collected nonwoven mesh. The path of electrospinning jet, being a combination of bending, looping and spiraling, generally leads to nanofibers in mostly a nonwoven web form with no preferred orientation. For different applications, several types of collectors have been developed such as plate type, drum type, disc type, wire mesh type and so on.

The plate collector is the oldest one, generally used for all initial studies on electrospinning in which the collector plate is placed at the opposite side of the spinning assembly. The resulting nanofibers have no preferred orientation.

There have been attempts to produce electrospun ultrafine fibers in an aligned fashion by the use of a cylinder surface moving linearly. The fibers are taken up tightly in a circumferential manner resulting in a fair alignment. The other route is to dampen the jet and control the deposition of electrospun fibers on a target. A series of charged rings acts as an electrostatic lens and changes the shape of the electric field. Modification in collector designs such as a moving planar surface [233] or a cylinder with a rotating high speed [234] and a collector with a gap [235] are other approaches to collect nanofibers in an aligned form.

An interesting and unique observation has been reported that electrostatic effects also influence the macroscale morphology of electrospun textiles and may result in the formation of heterogeneous or 3D structures [235]. It has also been observed that the residual

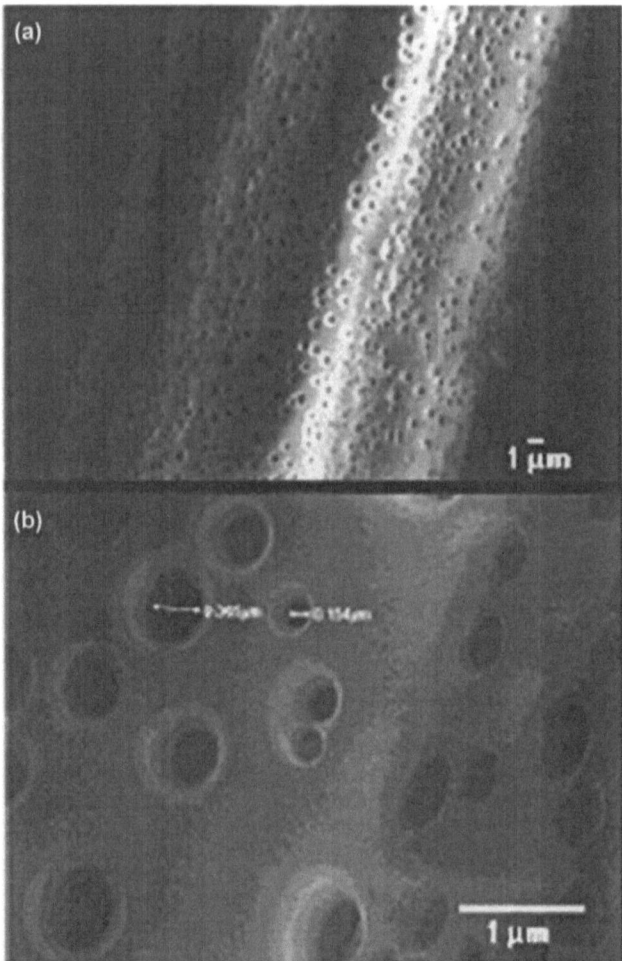

Figure 23.   Porous polystyrene electrospun fibers [221]. Reprinted with permission from S. Megelski, J.S. Stephans, D.B. Chase and J.F. Rabolt, Macromolecules 35 (2002) pp. 8456–8466. Copyright 2002, American Chemical Society.

charge present on electrospun fibers affects the way in which they organize themselves on nonwoven textile substrate. 3D honeycomb structures were observed by them under a microscope when using dilute solutions onto a wire screen. Megelski and coworkers [221] have created a highly micro- and nanostructured 'porous' morphology in an electrospun nanofiber (Figure 23), thus increasing their range of application significantly. The pores vary from densely packed, well-formed nanopores to large flat pores. The increased surface area of electrospun fibers was due to the highly volatile solvents used.

### 4.2.2.   Effect of process parameters

*4.2.2.1.   Effect of instrument variables.*   The voltage needed to eject a charged jet from the drop at the nozzle depends mainly on the solution viscosity. The threshold voltage to start the jet formation is plotted as a function of concentration. As concentration, or equivalently, the viscosity increases, higher electrical forces are required to overcome both the surface tension and the viscoelastic force for stretching the fiber.

Jet current is proportional to the transport of electrons, which is a measure of mass flow from the tip of the pipette to the grounded sheet. Results of experiments exhibit a power law dependence between flow rate and applied voltage, and the measured current and flow rate as:

$$\text{Flowrate} \sim (\text{Voltage})^3, \quad \text{Current} \sim (\text{Voltage})^{2.7}$$

Considering the similarity of the two scaling relationships, it can be concluded that the flow rate and current are linearly related. It is to be noted that the amount of polymer solution reaching the anode was usually less than the ejected amount of solution from the drop due to the various losses as the jet moved toward the anode [236].

Syringe-to-collector distance is one important variable for the spread and diameter of the nanofibers [237]. The jet diameter is decreased with an increase in distance from the spinning tip before it becomes unstable. Again, it has been found that a minimum distance is required between the tip and the collector to allow sufficient time for fibers to dry before reaching the collector. At distances that are either too close or too far, bead formation has been observed [238]. Buchko et al. [239] showed that with variations in the distance between the nozzle and the collector screen regardless of the concentration of the solution, lesser nozzle-collector distance produces wet fibers and beaded structures. Aqueous polymer solutions require more distance for the formation of dry fibers than systems using highly volatile organic solvents. However, for a number of polymers, there is no significant effect of the distance between the tip and the collector on the fiber diameter and morphology.

*4.2.2.2. Effect of salt addition.* Conductivity of the solution is the key factor in determining the spinning current. The addition of a small amount of salt was observed to dramatically increase the spinning current and subsequently the mass flow [236]. The ions increase the charge carrying capacity of the jet. It has been reported that by increasing the solution conductivity or charge density, more uniform fibers are produced with fewer beads [225]

*4.2.2.3. Effect of jet diameter.* The jet diameter becomes smaller as it travels to the ground owing to: (1) solvent evaporation, and (2) continuous stretching due to electrical force. The jet diameter is a function of the applied voltage. The diameter of the jet was determined by laser diffraction. The jet diameter seems to increase in a sigmoidal manner with the increasing voltage.

At high fields and low viscosities, more than one jet was ejected from the drop at the tip. Increase of the voltage favors the formation of several jets. The different fibers repel each other due to the flowing charge on their surface as a result of which the fibers distribute themselves at a larger area on the collector. Jets are observed to rotate in a clockwise direction in going from the nozzle to the collector. For the most concentrated solution of PU-urea at 12.8 wt%, a single jet was formed [236].

*4.2.2.4. Effect of solvent.* A study on preparation of nano-structured poly($\varepsilon$-caprolactone) (PCL) nonwoven mats by electrospinning process was carried out with three types of solution [240]. One was dissolved in only methylene chloride (MC), the second dissolved in a mixture of MC and *N*, *N*-dimethylformamide (DMF), and the third dissolved in a mixture of MC and toluene. MC, toluene and DMF are a good, poor, and nonsolvent for PCL, respectively. For the MC only, electrospun fibers had a diameter of about 5500 nm with a narrow diameter distribution, but for the mixture of MC and DMF, as the DMF volume fraction increased, the fiber diameter decreased to 200 nm. It was due to high electric properties of solution such as dielectric constant and conductivity. With an increase in toluene volume fraction, the electrospinning is strictly restricted due to a very high viscosity and a low conductivity. It can be interpreted that DMF has not only a high dielectric constant but also a polyelectrolyte behavior. The most interesting aspect of this study is the apparent change in the diameter and phenomenon of electrospinning observed in the electrospun PCL nonwoven mats, which strongly correlates to the dielectric constant of solution. It is clear that solution property is one of the important parameters in electrospinning.

*4.2.2.5. Nanofiber morphology: voltage dependence.* The electrospray process can be sustained in a variety of modes characterized by the shape of the surface from which the liquid jet originates. These modes occur at different voltages and have significant effects on droplet size distribution and current transport. Although distinct modes might be difficult to isolate and observe in the electrospinning process, it is expected that the degree of instability of the liquid surface from which the jet originates should produce changes in the electrospun fiber morphology. It has been verified experimentally that the shape of the initiating drop changes with spinning conditions (voltage, viscosity, feed rate). As the electrospray voltage is increased smoothly, the measured current undergoes stepwise increases that correspond to the observed changes in jet initiation modes [222].

In the case of electrospinning, the electric current due to the ionic conduction of charge in the polymer solution is usually assumed small enough to be negligible. The only mechanism of charge transport is the flow of polymer from the tip to the target. Thus, an increase in the electrospinning current generally reflects an increase in the mass flow rate from the capillary tip to the grounded target when all other variables (conductivity, dielectric constant and flow rate of solution to the capillary tip) are kept constant. Although there are no stepwise increases in the electrospinning current that would signal a change in mode, a change in electrospun fiber morphology may be correlated to the observed change in the slope of the electrospinning current as a function of voltage. The morphology of the electrospun nanofibers changes from that of primarily straight, defect-free fibers spun at a low initiating voltage to one in which the fiber mats contain a high density of beads when spun at a high initiating voltage. This bead structure becomes prevalent at a certain voltage, coincident with the change in slope in the plot of electrospinning current versus voltage. This coincidence suggests that by monitoring the spinning current, one may be able to control the bead defect density in the electrospun fibers [222].

The change in fiber morphology with voltage also correlates to changes in the originating droplet shapes. At low voltages, a droplet of solution remains suspended at the end of the syringe needle, and the fiber jet originates from a cone at the bottom of the droplet. The cone has a semi-vertical angle of about 50°, in agreement with Taylor's theoretical prediction of 49.3° for a viscous fluid in an electric field. The shape of this surface from which the polymer jet originates is similar in appearance to the microdripping jet mode in which the jet originates from the bottom of a drop whose diameter is larger than the capillary diameter. The nanofibers produced under these conditions have a cylindrical morphology with a few bead defects present. As the voltage is increased, the volume of the droplet decreases, the cone has receded and the jet originates from the liquid surface within the syringe tip. The electrospun fibers produced still have essentially a cylindrical morphology, but there is a distinct increase in the number of bead defects present in the fiber mat. At a still higher voltage, the solution jet appears to be initiating directly from the tip with no externally visible droplet or cone. At this voltage, the jet moves around the edge of the syringe tip, indicating that the jet originates on the inside surface of the syringe needle, where the edge of the liquid surface meets the needle wall. The fibers produced under these conditions have a high density of bead defects [222]. Bead formation has been initiated by varying solution surface tension and solution charge density, and by charge neutralization. The possibility that some small fluctuation in the charge density occurred as a result of charge dissipation from the tip into the atmosphere cannot be dismissed entirely. But corona discharge was not observed at voltages below 13 kV.

It is seen that electrospinning does not exhibit numerous discrete modes such as those observed in electrospray experiments. However, above a critical voltage, there are significant changes in the shape of the originating surface and the voltage dependence of the electrospinning current, and that these changes correspond to an increase in the number of bead defects observed in the electrospun fiber mats. The change in the shape of the liquid surface reflects a change in the mass balance that occurs at the end of the capillary tip. Increasing the voltage results in the rate at which the solution is removed from the capillary tip to exceed the rate of delivery of solution to the tip needed to maintain the conical shape of the surface. This shift in the mass balance results in a sustained but increasingly less stable jet. As the accelerating voltage is increased, electrospinning current generally increases step wise, which reflects an increase in mass flow rate from the capillary tip to the grounded target, when all other variables such as surface tension and charge density are kept constant. This is generally related to changes in originating droplet shape, which is related to bead defects in electrospun fibers. Fiber bead density increases with the increasing instability of the jet at the spinning tip and may be minimized through control of material flow rates to and from the tip [222]. So, at low voltages (just above the critical voltage), the electrospun fibers have a cylindrical geometry and a distinct decrease in bead defects; as voltage increases, the density of bead defects significantly increases.

*4.2.2.6. Nanofiber morphology: concentration dependence.* Solution surface tension and viscosity play an important role in determining the range of concentrations from which continuous fibers can be obtained in electrospinning. Solution concentration also affects the final size and distribution of nanofibers. At low viscosities ($\eta < 1$ poise), surface tension is the dominant influence on fiber morphology, and below a certain concentration, drops will form instead of fibers. At high concentrations ($\eta > 20$ poise), processing will be prohibited by an inability to control and maintain the flow of a polymer solution to the tip of the needle and by the cohesive nature of the high viscosity solutions. At low concentrations, the electrospinning process generated a mixture of fibers and droplets. The high viscosity solutions proved extremely difficult to force through the syringe needle of the apparatus used in these experiments, making the control of the solution flow rate to the tip unstable. For solutions with a very high concentration, the droplet at the end of the capillary stretched into a thick (diameter $\sim 0.5$ mm) strand that oscillated in the electric field. Eventually, the strand gained enough mass that gravity caused it to separate from the capillary and land on the ground plane. At no time was a sustainable jet achieved. Although the range of concentrations that produce fibers will obviously vary depending on the polymer/solvent system used, the forces of viscosity and surface tension will determine the upper and lower boundaries of processing window, if all other variables are held constant [222].

Even within the fiber processing window, varying solution concentration alters the morphology of the nanofibers formed. Two extremes in the fiber morphologies are observed as a result of varying solution concentration within the processable range. At the low concentration end of the processing window, the fibers have an irregular undulating morphology with large variations in diameter along a single fiber. There are numerous junctions and bundles of fibers. At high concentration end of the processing window, the nanofibers have a regular cylindrical morphology, and on average have a larger and more uniform diameter. At higher concentrations, the fibers exhibit a straight cylindrical morphology with relatively few fiber bundles and junctions, indicating that the fibers are mostly dry when they reach the collection screen. The apparent change in morphology may be reflective of the lower surface tension and solvent content in the high concentration solutions. In addition, measured mass deposition rates were found to decrease as the solution concentration increased. The effect of concentration on the average diameter of electrospun nanofibers is quantified and two basic observations can be made: first, the average diameter of electrospun fibers increases with spinning solution concentration, and second, the fiber population exhibits a bimodal size distribution above a certain concentration. A power law relationship was observed between fiber diameter and electrospinning solution concentration for electrospun polymer fibers, though the exponents differed. The differences in exponential power dependence may be attributed to the in-flight splitting or splaying of electrospinning jet. Several groups have reported splaying phenomena occurring during fiber production. However, only single-fiber production (the absence of splaying) in different polymer/solvent systems has been reported when electrospun under ambient conditions. The presence or absence of the splaying phenomena could result in different power law dependence for different polymer/solvent systems [222].

*4.2.2.7. Nanofiber morphology: temperature dependence.* Solution temperature is a key parameter that affects fiber morphology and spinnability. At room temperature, the highest concentration that yielded PU-urea fibers was 12.8 wt%. At high temperature, several jets were developed from the drop at the tip with a large angle between each other. Unlike the fibers generated from the 12.8 wt% solution at room temperature, fibers produced at a high temperature were surprisingly uniform in diameter, approximately 1 $\mu$m. It was also observed that the deposition rate increased as a function of temperature. Regardless of the applied electrical field, the deposition rate of fibers at high temperature is significantly higher as compared with that of fibers at room temperature, at equal deposition times. As a result, the film thickness (thickness of nonwoven fabric) is significantly affected by the solution temperature [236].

Further, Buer and coworkers have investigated the various electrospinning equipment design parameters such as controlled feed rate of polymer solution using a syringe pump, choice of capillary material, extraneous currents and varying collector geometries (static, moving screen or plate) and construction material, and so on. They observed that such trivial variables also have some influence on the quality of nanofibers produced [223].

### 4.2.3. Nanofiber properties

*4.2.3.1. Surface area.* Properties of the electrospun nanofibers, which may be relevant to their use in commercial applications, are discussed here. One property of the electrospun fiber textiles that is of interest is the total specific surface area associated with the textile. Electrospun fibers are recognized to possess high surface areas and large aspect ratios. The actual specific surface area of the electrospun textiles, obtained through Brunauer–Emmett–Teller (BET) measurements, ranged from 10 to 20 m$^2$ g$^{-1}$. These values are several orders of magnitude higher than one would expect from fabrics composed of ordinary textile fibers, whose diameters are typically in the order of 10–100 $\mu$m. The BET measurements were not accurate enough to evaluate subtle changes in specific surface area as a function of fiber size changes induced by changes in electrospinning solution concentration.

*4.2.3.2. Crystallinity.* The crystalline properties of the electrospun fibers may also be of primary importance when considering the materials for commercial applications. WAXD and DSC experiments on samples of PEO nanofiber textiles show WAXD patterns for both a PEO nanofiber textile spun from a 10 wt% concentration PEO/water solution and the neat PEO powder from which the solution was made. The peak positions in each diffraction pattern are essentially identical, showing that there is no change in crystal structure induced by the electrospinning process. However, it is immediately evident that the diffraction peaks associated with the PEO powder are sharp and clearly defined, while the peaks in the pattern from the nonwoven textile are significantly broader, indicating that the crystalline order in the nanofibers was significantly less than in the powders. Similarly, the PEO nanofiber textiles had a melting temperature, $T_m$, and heat of fusion, $\Delta H_f$, (66°C and 200 J/g, respectively) that were much lower than $T_m$ and $\Delta H_f$ found for the neat powder (71°C and 250 J/g, respectively), indicating that the overall crystallinity of the electrospun fibers was poor [222]. These results are in qualitative agreement with the results presented by Larrondo in which relatively low crystallinity, spherulitic structure and unorientation were observed for polymer fibers electrospun from the melt state [226]. The chains and crystallites in nanofiber have some orientation to the same direction of fiber axis during electrospinning process. Also, the retardation of crystallization has been observed in electrospun PCL nonwoven mats [236].

*4.2.3.3. Residual charge.* A number of observations suggest that under certain conditions, the residual charge left on the electrospun fibers affects the way in which they will organize themselves in the nonwoven textile. Specifically, the fibers will try and arrange themselves in such a way as to maximize contact with the electrically grounded target. For example, visual inspection of any electrospun textile that has been collected on a metal screen will reveal that it is thicker at points in contact with the wires. Attempts to electrospin from solutions with low viscosities have resulted in the collection of electrospun fibers in geometries other than a two-dimensional fiber mat or a film. Instead of a uniform nonwoven fiber textile, the material can be collected preferentially on the wires ultimately building up a 3D cellular structure on the grid. 3D fiber structures have also been observed in fiber mats collected on other types of substrates. For example, a honeycomb structure was electrospun at a high voltage onto a piece of cloth backed by an aluminum screen. Because of the high electrospinning voltage, the spinning rate is faster and the electrospun fibers are encountered at the cloth substrate wet and full of bead defects, readily identifiable at a high magnification. These beads and fibers adhere to each other forming a thick, continuous layer of polymer particles that eventually builds up enough electrostatic charge to prevent new electrospun fibers from laying directly on the mat. Instead, they are suspended slightly above the surface, where they have time to dry, forming the 3D honeycomb structure.

All the above examples show that the electrospun polymer that reaches an electrically grounded target can carry enough residual charge to influence the morphology of the textile formed. Although it is possible to eliminate such structure by judicious choice of

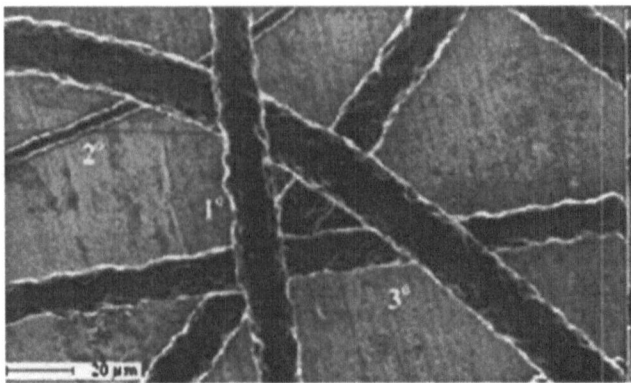

Figure 24. Electron micrographs of size distribution in average fiber diameter. Fibers obtained from electrospinning of 12.8 wt% PU-urea belonging to primary, secondary and tertiary populations are demonstrated [236]. Reprinted from M.M. Demir, I. Yilgor, E. Yilgor and B. Erman, Polymer 43 (2002) pp. 3303–3309; Polymer 42 (2001) pp. 9955–9967, with permission from Elsevier.

processing parameters, for certain applications, the 3D networks of submicron fibers formed by electrostatic effects may be desirable [222].

*4.2.3.4. Size and size distribution.* Electrospinning is thus a fast and simple process of producing a wide range of nanofibers. The collected webs of nanofibers usually contain fiber with varying diameters from 50 nm to 2 microns. Sometimes, the main fiber breaks up into many smaller filaments, a process called 'splaying', which happens because of electrostatic instabilities under certain conditions. Compared with melt blown nanofibers, electrospun fibers have a much narrower diameter distribution. Electrospun fibers were not uniform in diameter and morphology. Both concentration and temperature have significant roles in determining the fiber size distribution and morphology. The three different sized fibers are denoted as primary (1°), secondary (2°) and tertiary (3°) in Figure 24. The diameter of the primary population of fibers is approximately 1 $\mu$m. The second population is nearly one-third of the primary one, which is approximately 0.4 $\mu$m, whereas the tertiary population is approximately 1.4 $\mu$m in diameter.

A bimodal distribution has been reported to occur in electrospinning of 10 wt% PEO in aqueous solution. The distribution of diameters observed in electrospinning of 12.8 wt% PU-urea fibers is shown in Figure 25. In addition to the primary and the secondary peaks noted on the curve, a small tertiary peak is also observed. The size of the primary population is larger than the total size of the secondary and tertiary population [236].

*4.2.3.5. Morphology.* Fiber morphology of PU-urea fibers also varies with the concentration of solution subjected to electrospinning. Different fiber morphologies occur at different concentrations and have significant effects on surface area to volume ratio of the fibers. At high concentrations, fibers exhibit curly, wavy and straight structures. The nanofibers as shown in Figure 24 have identical knots at regular distance. In addition to the curly structures, wavy and straight structures are also observed on the same sample. On the contrary, fibers obtained from low-concentration (5.2 wt%) solutions exhibit 'beads on string' morphology. The average bead length was 700 nm. Lower viscosity solution favors the

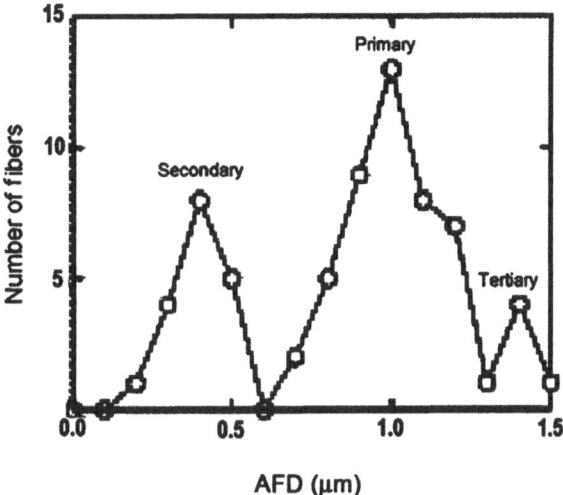

Figure 25. Diameter distribution of nanofibers obtained from 12.8 wt% concentrated PU-urea solution [236]. Reprinted from M.M. Demir, I. Yilgor, E. Yilgor and B. Erman, Polymer 43 (2002) pp. 3303–3309; Polymer 42 (2001) pp. 9955–9967, with permission from Elsevier.

formation of beads and also the formation of thinner fibers. Beads are known as defect structures because they disturb the unique property of electrospun fibers and decrease the surface area to volume ratio. The occurrence of the bead formation stems mainly from the high electrical field applied to the system. Increasing the distance or decreasing the electrical field decreases the bead density. The nanofibers in the first image were generated at an electrical field of 2.35 kV cm$^{-1}$, whereas nanofibers in the second image were obtained at 0.52 kV cm$^{-1}$. To this end, the morphology of electrospun PU-urea fibers changes from curled, at high concentration, to one containing beads, at low concentration. Both morphological properties have a negative influence on surface area to volume ratio of electrospun fibers. It is, therefore, desirable to generate fibers without beads or curled structures, in other words without any 'by products'. Entov has reported that the formation of beaded fibers is related to the instability of the jet of polymer solution. Fong has also reported about beaded PEO nanofibers. It has been observed that the viscosity, net charge density and surface tension of solution are key parameters to form beaded fibers [236].

The electrospun fibers are one or two orders of magnitude smaller in diameter than wet-spun monofilament, but they are in the form of nonwoven fabric. The electrospun fibers are excellent candidates for membrane technology as the chaotic deposition makes a nanoscale porous structure. However, the lack of orientation in this form of collection of electrospun fibers hinders them to form yarn. The wet-spun fibers are known to have a high degree of surface roughness as compared with fibers spun by other conventional techniques. AFM measurements of the surface roughness of the electrospun fibers show that the latter are more than two times rougher than the wet-spun fiber of the same polymer. The roughness measured by the AFM depends on the size of the area selected [236].

Trifluoro Graft Elastomer (TrF1) is a copolymer in which the grafted segments are made up of piezoelectric polymer. Electrospinning of solutions of this material under a variety of conditions demonstrates the effect that spinning parameters have on the final arrangement and quality of the fibers. A typical optical microscope image of this electrospun polymer showed more globular masses than fibrous material. At a lower magnification, a wet coating was observed, indicating prevalence of polymer spraying versus fiber spinning. However, at a higher magnification, some fiber deposition was evident among the polymer droplets. Additionally, agglomeration was apparent at some junction points of the fibers; this feature demonstrates that the fibers were not completely dry upon deposition. However, minor changes in any of the spinning parameters yielded widely different mat quality.

TrF1 fibers covered the wing frame forming a mat and remained securely attached and intact when the frame was removed from the collector. The resulting fibrous coatings exhibited a degree of strength and elasticity consistent with the small diameter of the individual fibers and the complex network that the fibers created. The electrospun fibrous mats had an expected property knockdown compared to the film, which exhibited a stiffness of 86 ± 1 MPa and an elongation of 985 ± 29%. The electrospun fibers created an unoriented mesh, with bonded contact points between fibers. Upon elongation, contact points in the mat were broken, and finally, fibers were broken. As the mats were pulled in tension, contributions from mesh properties as well as fiber properties were observed. The fiber-coated wing frames exhibited a perceptible vibration upon excitation with a 2 kV (peak-to-peak) sine wave at 6.7 Hz [241].

The effect of linear velocity of collector drum surface on tensile properties of electrospun nanofiber is significant. It has been found from the stress–strain curves of electrospun PCL nonwoven mats that they exchanged mechanical properties of cross-direction and machine direction as linear velocity of the drum surface increased. For low velocity (1.3 m min$^{-1}$), tensile strength and modulus in the cross-direction were higher, but at medium velocity (3.2 m min$^{-1}$), these were higher in machine direction. But elongation at break was not changed with velocity. For high linear velocity of drum surface (>4.5 m min$^{-1}$), all of the properties decreased. To explain the results, two parameters need to be considered: fiber orientation and bonding point between fibers. Electrospun fiber orientation can be described using a fiber orientation angle, which is the angle formed between the fiber axis and a line parallel to the web centerline. Also, the formation of point bonding structures affects physical and mechanical properties of nonwoven mats [236].

*4.2.3.6. Cross-sectional shape.* In addition to round nanofibers, electrospinning a polymer solution can produce thin fibers with a variety of cross-sectional shapes. Branched fibers, flat ribbons, ribbons with other shapes and fibers that were split longitudinally from larger fibers were observed. The transverse dimensions of these asymmetric fibers were typically 1 or 2 $\mu$m, measured in the widest direction. The observation of fibers with these cross-sectional shapes from a number of different kinds of polymers and solvents indicates that fluid mechanical effects, electrical charge carried with the jet and evaporation of the solvent all contributed to the formation of the fibers. The influence of a skin on the jets of polymer solutions accounts for a number of the observations [242].

*4.2.4. Melt spinning of nanofibers*

The use of molten polymers to produce electrospun mats is a subject of great interest. Nanofibers of polymers can be electrospun by creating an electrically charged jet of polymer solution or melt at a pendent droplet. Although solution spinning is most popular for electrospinning, several researchers also concentrate on the melt spinning possibility of electrospinning. The main problem with electrospinning is that the rate of fiber production is extremely low. Methods for scaling up production have not yet been particularly success-ful. Melt electrospinning can be scaled up to provide significant increases in productivity. In melt electrospinning, polymer melt is ejected from the capillary tip. The fiber will cool and solidify as it is drawn toward the collector. In spite of the potential benefits of melt electrospinning, little progress has been made in the past 20 years. Larrondo and Manley [226,227,229] were the first to electrospin a molten polymer more than two decades ago. They were capable of spinning PP (melt flow indexes 0.5–2.0) and succeeded in making fibers that were greater than 50 $\mu$m in diameter. Their inability to spin submicron fiber was attributed to the large increase in viscosity that could be many orders of magnitude greater than that of a polymer solution. They observed that the polymer melt experienced a large initial decrease in diameter when placed in an electric field, and as the electric field strength increases, the fiber diameter decreases. However, there is a lack of bending insta-bility that hinders the production of submicron fibers. This is attributed to the extremely high viscosity associated with the melt, as has been demonstrated by Taylor, or the small distance that the jet traverses before contacting the collection device. The solidified fibers are deposited randomly on the surface of the grounded collection plate. Other groups have conducted research on melt electrospinning polymers including poly(ethylene terephtha-late) and polyethylene. These groups reported a wide range of obtainable fiber diameters, yet limited progress has been made [243].

The key to spin nanofibers in melt spinning route is the control over temperature and air flow in the spinning chamber. Nanofibers and small microfibers have been obtained by delaying quenching of the fiber extrudate [244]. But to form consistent high quality fibers in the nano-meter range, precise control over the fiber temperature and supramolecular structure throughout the process must be achieved. Because the fibers are so small in diameter, a simple thermal analysis reveals that they will almost instantaneously reach the temperature of their environment. If the temperature is controlled to prevent the cooling, the melt behaves in a similar manner as the solutions under an electric field, as it flows away from the droplet in a nearly straight line, it bends into a complex path and other changes in shape occur, during which electrical forces stretch and thin it by very large ratios.

Control over the fiber temperature requires a precise control over the temperature of the air surrounding the fiber. The challenge of creating a device that provides precise temperature control is made somewhat more difficult by the necessity of avoiding the use of any electrically conductive materials, as they interfere with the electric field necessary to draw the fiber. The air currents also need to be precisely controlled as nanofibers are extremely sensitive to air flow and likely to be affected by the buoyant effects of heated air.

It has been shown that fiber diameter can be controlled by adjusting the process-ing parameters such as electric field strength, polymeric viscosity and flow rate. A full understanding of the melt electrospinning process, and its potential to replace solution electrospinning, has not yet been realized [243].

### 4.2.5. *Nanofibers from a variety of polymers*

A variety of polymers have been used to produce nanofibers through the electrospinning route. Theoretically, any melt or solution spun fiber can be spun into a nanofiber through this process as all of them have moderate electrical conductivity and high viscosity. The ranges of fibers produced through electrospinning are acetate rayon, acrylic, polyolefin (from melt), PEO, polybenzimidazole, PGA, PU, polyaramids, nylon 6, polyvinyl alcohol and so on. A review on polymer nanofibers by Huang et al. listed 50 types of electrospinnable polymers among which six are melt spinnable polymers [220].

Electrospinning of poly(ethylene-co-vinyl alcohol) or EVOH copolymers from 2-propanol-water solutions is a straightforward and general route for the production of highly fibrous and porous EVOH materials for various biomedical applications. EVOH, ranging in composition from 56 to 71 wt% vinyl alcohol, can be readily electrospun at room temperature from solutions in 70% 2-propanol/water (rubbing alcohol). The solutions are prepared at 80°C and allowed to cool to room temperature. Fiber diameters of ca. 0.2–8.0 $\mu$m were obtained depending upon the solution concentration, an attractive range for tissue engineering, wound healing and related applications. Electrospun EVOH mats have been shown to support the culturing of smooth muscle cells and fibroblasts [245].

Aliphatic polyesters such as poly(lactic acid) (PLA), poly(glycolic acid) (PGA) and PCL and their copolymers are biodegradable polymers that have been used in biomedical and industrial application due to their biodegradability. PCL has crystallizable rubbery properties, so it has been widely utilized for improving elasticity. MONOCRYL is a commercial monofilament suture that was formed from a segmented block coploymer of glycolide and caprolactone. Development of electrospinning has been rapidly increasing in the past few years because it can prepare fibers that are smaller than the diameter of conventional fibers by 100 times [236].

The ultrafine fibers have been prepared by electrospinning a poly(amic acid) solution, a precursor of polyimide, followed by thermal imidization. The fiber diameters, which are much smaller than conventionally spun fibers, range from a few tens of nanometers to several micrometers. A rectangular cross-section is observed in the case of submicron fibers, with a cross-sectional dimension below approximately 500 nm [246].

Nonwoven nanofibers of *Bombyx mori* and *Samia cynthia ricini* silk fibroins, and of the recombinant hybrid fiber were prepared involving the crystalline domain of *B. mori* silk and noncrystalline domain of *S. c. ricini* silk from hexafluoroacetone (HFA) solution using the electrospinning method. A high electric potential is applied to a droplet of a silk solution at the tip of a plastic capillary in which a Pt wire is used as an electrode. The jet dries as the solvent is evaporated in the air. The dried fibers are collected on a receiving conducting mesh. The collecting mesh was placed at distance of 10–15 cm from the capillary tip. Additionally, the capillary axis had to be tilted about 5–10° from the horizontal direction. A voltage of 15–30 kV was applied to the wire in the capillary through a high voltage power supply, while the receiving mesh was grounded. Methanol was used as a solvent to remove HFA from the electronspun fiber and induce the structural change in the post-spin stage. The nonwoven fibers were soaked in a methanol bath overnight to allow HFA to diffuse from the fiber, and then the nonwoven fibers were dried in the vacuum overnight at room temperature [247].

Poly(vinyl alcohol) (PVA) is a polymer that has been studied intensively because of its good film-forming and physical properties, high hydrophilicity, processability, biocompatibility and good chemical resistance. These properties have led to its broad industrial use in areas such as membrane, textile sizing and finishing, adhesive, and coatings and paints. PVA properties can be improved or modified by importing other composites. Therefore, some PVA/silica nanocomposite materials have been reported. However, among them, only the gel or film properties of PVA/silica nanocomposite materials were investigated. Fiber mats of PVA/silica composite with different silica content were successfully prepared by the electrospinning technique. These might be the first fiber mats of organic–inorganic composite materials, which would give a new state and some special properties of the organic–inorganic composite materials distinguished from the state of film and gels [248].

To develop the wing skin of micro air vehicles (MAV), it was desirable to have a tough, durable and easily processable electroactive polymer (EAP), polymer capable of reacting quickly with high strain and mechanical output. A piezoelectric copolymer of poly(vinylidene fluoride) and trifluoroethylene was found to be most suitable, as this copolymer offers a higher crystallinity than PVDF alone, and it is not necessary to stretch the material to induce a polar crystalline phase. This copolymer can be dissolved in dimethyl formamide (DMF) solvent and electrospinning of this copolymer on the wing frame of MAV leads to fabrication of lightweight, responsive MAV wings [241].

### 4.2.6. Functionalized nanofibers

Functionalized nanofibers can be defined as nanofibers with specific additives for imparting special functionalities and capabilities to nanofibers, thus widening the scope of their applications. The materials that have been added are metal/metal oxides at a nanorange, biological materials such as enzymes, drugs, CNTs (SWNTs as well as MWNTs) as well as nanoclays [249]. These value-added nanofibers can be used effectively for several high-end applications in filtration, chemical protective clothing, biomaterials, drug delivery and tissue engineering. These polymeric nanofibers can also find applications in chemical and process industries such as catalysts, physical and chemical adsorption processes, and so on, and are excellently reviewed by Ramkumar et al. [250].

Recently, electrospinning has also been extended to making nanofibers from polymer nanocomposites, incorporating nanoclays, CNTs and other nanoparticles and adding a new dimension to nanofibers. Fong et al. [251] were the first to report electrospun nanofibers (~100 nm) based on nylon 6/exfoliated MMT clay using hexafluoro-iso-propanol (HFIP) as the solvent. The resulting nanofibers had highly aligned MMT layers (normal to fiber axis) and nylon 6 crystallites (layer normal/parallel to fiber axis). Other polyamide 6,6 or polyamide nanocomposite electrospun fibers have also been reported based on MMT clay [252] as well as CNT as nanofiller [253]. Polyester/CNT composite nanofibers have been produced through electrospinning at varying CNT content (0.001–90 wt%) [254]. These composite fibers can be further used to manufacturer fabrics, antistatic materials, electromagnetic shielding materials, high performance separation medium, reinforcing materials, electrical and thermal conductivity materials, wave absorbing materials and so on.

Fibers and nanofibers of PA 6/MMT clay nanocomposites (diameter between 100 and 500 nm) can be collected as nonwoven fabrics or as aligned yarns. The electrospinning process resulted in highly aligned MMT layers (layer normally perpendicular to the fiber axis) and Nylon 6 crystallities (layer normal, parallel to fiber axis) [252,255]. The PA 6/clay nanofiber web deposited on a nonwoven substrate as a continuous coating increased the contact angle and the time required for water penetration relative to the uncoated substrate [255].

The parameters that have to be carefully monitored are polymer and solvent selection, electrical field control, solvent evaporation and management and nanofiber web formation and characterization. Quality control of nanofiber web also requires novel techniques based on real-time measurement about web consistency.

### 4.2.7. Nanofiber applications

Electrospun nonwoven mats have small pore size, high porosity and high surface area; therefore, they can be used in a wide variety of applications such as reinforcing fibers in

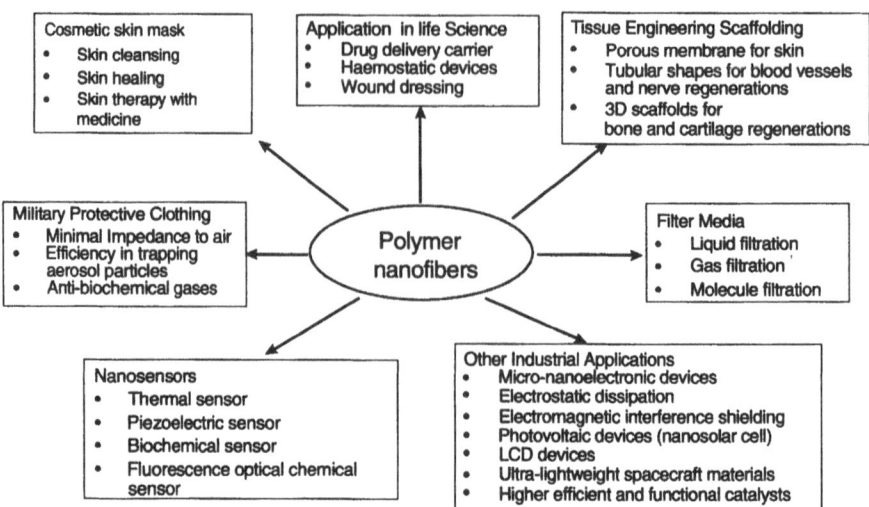

Figure 26. Potential applications of electrospun nanofibers [220]. Reprinted from Z.M. Huang, Y.Z. Zhang, M. Kotaki and S. Ramakrishna, Comp. Sci. Tech. 63 (2003) pp. 2223–2253; Polymer 42 (2001) pp. 9955–9967, with permission from Elsevier.

composite materials and for scaffolds in tissue engineering, and so on [236]. Because of their larger surface to volume ratios, they are able to absorb more liquids than fibers having large diameters. Small pore sizes of electrospun fibers make them suitable candidates for military and civilian filtration applications. The fibers having smaller diameter have lower bending moduli resulting in softer fabric hand. These fibers are attracting considerable interest in a wide range of applications, including filters, membranes, tissue scaffolding, composites, biomimetic materials and nanoelectronics. The potential areas of applications are described in Figure 26.

The morphological studies indicate that the electrospinning process does partially orient the molecules in fibers, although to date all nanofibers produced are without any control over forces during orientation and crystallization. Nanofiber technology has not yet developed commercially, and therefore, engineers and entrepreneurs have not had a source of nanofiber to incorporate into their designs. Uses for nanofibers will grow with improved prospects for cost-efficient manufacturing, and development of significant markets for nanofibers is almost certain in the next few years. The leaders in the introduction of nanofibers into useful products are already underway in the high-performance filter industry. In the biomaterials area, there is a strong industrial interest in the development of structures to support living cells. The protective clothing and textile applications of nanofibers are of interest to the designers of sports wear, and to the military, as the high surface area per unit mass of nanofibers can provide a fairly comfortable garment with a useful level of protection against chemical and biological warfare agents.

Donaldson Company Inc., USA, is one of the pioneers in the field of nanofibers and has commercialized the production of nanofiber webs from electrospinning for a broad range of filtration applications [256]. Nanofibers provide dramatic increases in filtration efficiency at relatively small decreases in permeability. Nanofiber filter media can contain airborne contamination in the personal cabin of mining workers. However, the production process and quality control of electrospinning nanofiber webs present special challenges.

CNFs are potentially useful in reinforced composites as supports for catalysts in high temperature reactions, heat management, reinforcement of elastomers, filters for liquids and gases, and as a component of protective clothing. Nanofibers of carbon or polymer are likely to find applications in reinforced composites, substrates for enzymes and catalysts, applying pesticides to plants, textiles with improved comfort and protection, advanced filters for aerosols or particles with nanometer scale dimensions, aerospace thermal management application and sensors with fast response times to changes in temperature and chemical environment. Ceramic nanofibers made from polymeric intermediates are likely to be useful as catalyst supports, reinforcing fibers for use at high temperatures, and for the construction of filters for hot, reactive gases and liquids.

Most of the recent work on electrospinning has focused either on trying to understand the fundamental aspects of the process in order to gain control of fiber morphology and structure or on determining appropriate conditions for electrospinning of various polymers. Given that most proposed applications exploit the high surface area of the fibers and textiles, methods for tailoring the chemistry of the electrospun fiber surface should be important to the eventual exploitation of the technology. Successful engineering of surface chemistry in electrospun nanofibers may be exploited directly in the design of membranes and demonstrates a pathway for surface functionalization of electrospun fibers for biomedical, composite reinforcement and other applications [257].

A new class of materials based on organic and inorganic species combined at a nanoscale has attracted more attention recently. These new materials, called nanocomposites or organic–inorganic hybrids, have the possibility to become new materials having both the advantages of the organic materials such as lightweight, flexibility and good moldability, and of inorganic materials such as high strength, heat-stability and chemical resistance. These composites are expected to be applied for scratch- and abrasive-resistant-hard coating, nonlinear optical materials, contact lenses, and reinforcement of elastomers and plastics. In composite applications, if there is a difference in refractive index between fiber and matrix, the resulting composite becomes opaque or nontransparent due to light scattering. One possible way of circumventing this limitation would be to use fibers with a diameter significantly smaller than the wavelength of visible light. Other excellent examples are thin fibers for filter applications, fiber mats serving as reinforcing component in composite system, biomedical applications and template for the preparation of functional nanotubes. As a result, making polymer/inorganic composite extra thin fibers will combine the advantages of organic–inorganic hybrids and thin fibers together while retaining the transparency of the composites, and this would meet the need of new materials [248].

Processing of EAPs via electrospinning introduces the potential for designing an active wing that can be tailored to achieve flight adjustments such as turns and elevation changes in MAV applications. MAVs are small, lightweight vehicles that weigh less than 50 g, valued for their versatility and mobility. They are useful for collecting and transmitting visual images from hazardous locations. Most MAV designs rely on scaled down versions of conventional fixed wing aircraft, incorporating a flexible wing structure consisting of a graphite fiber composite frame covered with a latex skin. An optimal polymer composition was electrospun onto MAV wing frames to create a bird wing-like texture that will permit mimicry of the agility and versatility in control exhibited by birds in flight. The results confirm the feasibility of utilizing electrospinning to process novel EAPs for the fabrication of lightweight, responsive MAV wings [241].

Some of the commercially available nanofibers for different applications available in the market are shown in Table 3.

Table 3. Commercially available nanofiber-based products.

| Sl. No. | Company | Product name | Type of product | References |
|---|---|---|---|---|
| 1. | Hollingsworth & Vose | Nanoweb | Air and liquid filter. | www.hollingsworth-vose.com |
| 2. | Mempro | Filterhot | Catalyzed ceramic nanofiber air filter, used to reduce pollution in coal fuel power plants. | www.mempro.com |
| 3. | AMSOIL | AMSOIL Ea oil filters | Oil filters with advanced full synthetic nanofiber technology, making them the highest efficiency filters that are available for the auto/light truck market. | www.amsoil.com |
| 4. | Stealth Audio Cable | Nanofiber (pure carbon) | Analog interconnects based on pure carbon nanofiber. | www.stealthaudiocables.com |
| 5. | Beauty Cloth International | Beauty Cloth Exfolia™ facial cloth | The nanoscale fibers remove the dead scale of the skin without affecting the lower layers. | www.beautycloth.com/exfolia/ |
| 6. | Donaldson Company Inc. USA | Industrial filters | Air, oil and hydraulic nanofiber-based filters. | www.donaldson.com |
| 7. | Cosmo Bio Co. Ltd. | Chitosan nanofiber matrices | Cell and tissue culture substrates based on chitosan nanofiber. | www.cosmobio.co.jp |
| 8. | Evouni Corporation | iPhone Nano Fiber Pouch | Protecting iPhone from dusts, iPhone can comfortably lay on Nano Fiber Pouch as desk pad. | www.tootoo.com/s-ps/iphone-nano-fiber-pouch-p-1294356.html |
| 9. | eSpin Technologies | SIMWYPES™ | Super dusting cloths, capable of removing invisible hazardous chemicals. | espin.karmonfrench.com |
| 10. | Teijin Fibers Ltd. | NANOFRONT™ | Ultrafine polyester nanofiber, suitable for a variety of applications, including functional sportswear, innerwear, skin care products, antibacterial filter, precision grinding cloth, etc. | www.teijinfiber.com/english |

### 4.2.8. *Other techniques to produce nanofibers*

Electrospinning techniques, however, have been problematic because some spinnable fluids are very viscous and require higher forces than electric fields can supply before sparking occurs, that is, there is a dielectric breakdown in the air. Likewise, these techniques have been problematic where higher temperatures are required because high temperatures increase the conductivity of structural parts and complicate the control of high electrical fields. Combining electrospinning techniques with melt-blowing techniques is creating some interest. But the combination of an electric field has not proved to be successful in producing nanofibers in as much as an electric field does not produce stretching forces large enough to draw the fibers because the electric fields are limited by the dielectric breakdown strength of air.

Although electrospinning has been most widely reported in producing nanofibers, the low production rates of this process are a major limitation in up scaling this technique, and thus, make it mostly a laboratory curiosity. The production rate of nanofibers through electrospinning is generally measured in grams per hour, which is essentially very low by a single needle. So, attempts have been made to speed up the process using multiple syringes [258] as well as using needleless spinning. In needleless electrospinning, several jets are coming out from the polymer dope surface under a strong electric field [259]. Yarin et al. used a magnetic liquid below the polymer dope to produce a number of spikes in which the electrical fields are concentrated to produce a huge number of jets [260].

The commercial success for industrial-scale production of nanofibers is Nanospider, a needleless electrospinning system from Elmarco S.R.O, which produces uniform nanofiber webs continuously from free liquid surface at a very high production rate (http://www.elmarco.cz/technology/nanospider%3Csup%3Etm%3Csup%3E-technology/). Nanofibre Inc. Aberdeen, North Carolina, USA, has developed a unique process of producing nanofibers through melt blowing using a molecular dye [261]. The fibers produced are a mixture of both micron and submicron-sized fibers. But it is a relatively inexpensive technique and can produce nanofibers in large quantities at a cost of $10 per kg. The concerns are a broad range of fiber diameters produced. If perfected, this technique can certainly take the nanofiber production to a commercial future because of higher production rates as compared with electrospinning.

Another technique is based on making bicomponent fibers that split or dissolve. The most researched is the production of islands in the sea fiber using a standard spin/draw process. The production rate is 5 kg/h at a take speed of 2500 m$^{-1}$ min. Both PET and PA 6 nanofibers of diameters approximately 300 nm can be produced using EVOH as a sea polymer, and the ratio being 50/50. Unlike electrospinning and melt blown technique, the nanofibers produced are in a narrow range and the projected cost is 1–$5 per kg. Another possible approach is the use of bicomponent fiber spinning to manufacture nanofibers by the splitting process. The number of segments needs to be greater than 16 and a water-soluble polymer should be used in a small ratio along with PET and PA 6 [262]. Toray industries Inc. (http://www.just-style.com/companies/toray-industries-inc_id462) has developed a new technology for producing multifilament nanofibers comprising monofilaments whose diameter is in the nanorange. The technology is versatile not only in its ability to be applied to commodity polymers such as nylon or polyester but also in that nanofibers can be manufactured using existing productions equipments. They have succeeded in producing for the first time nylon nanofibers having uniform diameter in nanorange through optimization of rheological properties of polymers themselves. As nylon nanofibers have a multifilament continuous structure, they can be easily fabricated into a wide array of products while the orientation and shape of the nanofibers can be easily controlled, enabling their application to be extended into various fields.

The nylon nanofibers produced have about 1000 times larger surface area than the conventional fibers; thus, adsorption and adhesion properties and novel functions attributable to the fiber surface are also more pronounced and promise a variety of applications as advanced materials. The moisture absorption of these nylon nanofibers is about two to three times higher than conventional nylon, which is comparable to cotton.

In another ongoing NTC project at Georgia Tech, USA [263], a unique method of producing nylon and polyester nanofibers using 'extrusion polymerization' through mesoporous silica and aluminum channels as nanoreactors is being investigated. Unlike electrospun nanofibers, which are neither strong nor stiff because polymer molecules in these fibers are not oriented, the nanofibers produced using this new brilliant technique will possesses high mechanical properties as they have a high degree of crystallinity from extended chain crystals formed inside the nanochannels.

Thus, new researches in producing nanofibers in a continuous and aligned mono- or multi-filament form with more uniform diameter and narrow range have a potential to bring nanofibers in the market place, at a reasonable price and acceptable production rates. Nanofibers have been produced mostly through electrospinning from a wide range of polymeric materials [264], but other techniques as described above [265] also show much promise in the future success of nanofibers.

## 5. Conclusion

The enhancement of several attributes of textile materials via nanotechnology is expected to become a huge industry in the next decade, with tremendous technological, economic and ecologic benefits. Nanotechnology is drawing the attention of entrepreneurs and investors for its potential to make substantial improvements in the overall quality of life. Most of the nanotechnology-based products, currently in their developmental stage, are expected to be launched in the market within the next few years. The nanotechnology industry is currently in the late introduction or early growth phase. As the nanotechnology products are initially expected to be more expensive than the traditional products, convincing the customer about the value addition against the additional cost is essential for the growth of this technology. The use of nanoparticles, nanofinishes and nanostructures is capable of developing and enhancing advanced performance characteristics of conventional textiles, in areas such as anti-microbial, anti-bacterial, water repellency, soil-resistance, anti-static, anti-infrared and flame-retardant properties. The dyeability and color fastness of textile materials can also be enhanced by creating nano structural surfaces. The technology can be used in engineering desired textile attributes, such as strength, softness, durability, comfort and breathability in fibers, yarns and fabrics. For the new generation of nanocomposite fibers and coatings, the key issue lies in uniform dispersion of nanoparticles, strengthening of the nanoparticle agglomerates and enhancement of nanoparticles/matrix interaction. Homogeneous dispersion of nanoparticles in polymers is very difficult because the particles possessing high surface energy have a tendency to agglomerate. Consequently, a number of loose clusters of nanoparticles appear in matrices and the nanocomposites exhibit properties even worse than conventional micro-particle/polymer systems. Inorganic nanoparticles are usually modified to improve their compatibility with polymer matrix and achieve a homogeneous dispersion in polymer matrix as well as improved properties of polymer nanocomposites. The future of high performance commodity textiles depends on nanocomposite fibers that have the potential to meet the growing customer demands on the functional performances of textiles. The electrospinning process along with different winding technologies for aligning the nonwoven nanofiber mesh produced have their tremendous potential for high- end-smart, multifunctional textile applications. The strength of these fibers and the yield is still an issue which is expected to be overcome soon.

There is a significant potential for profitable applications of nano technology in cotton and other textiles. In future, interdisciplinary research collaborations will lead to significant advancements in the desirable attributes of textiles for a variety of applications that include healthcare, protective clothing, filtration media, biomedical engineering and so on. The textile industry has the biggest customer base in the world. Therefore, advances in the customer-oriented products should be the focus for the future nanotechnology-based textile applications. The future research should also be targeted on developing medical and healthcare textiles, improved dirt, crease and shrink resistance properties of fabrics, temperature adaptable clothing and odorless undergarments. The toxicological issues of handling and applying various nanoparticles also have to be closely watched and monitored as they find increasing applications in commodity products including textiles.

## References

[1] T. Matsuo, Text. Prog. 40 (2008) pp. 123–181.
[2] Q.Q. Zhao, A. Boxman and U. Chowdhry, J. Nanopart. Res. 5 (2003) pp. 567–572.
[3] K.A. Padmanabhan, Mater. Sci. Eng. A 304–306 (2001) pp. 200–205.
[4] K.T. Lau, M. Chipara, H.Y. Ling and D. Hui, Composites Part B 35 (2004) pp. 95–101.
[5] M. Jaffe, G. Collins and J. Menczel, Thermochim. Acta 442 (2006) pp. 95–99.
[6] K.P. Chong, J. Phys. Chem. Solids 65 (2004) pp. 1501–1506.
[7] S. Guceri, Y.G. Gogotsi and V. Kuznetsov (eds.), *Nanoengineered Nanofibrous Materials*, Kluwer, New York, 2004, pp. 245–468.
[8] S. Mukhopadhyay, Text. Rev. October (2007) pp. 85–99.
[9] L. Qian, J. Text. Apparel Tech. Manag. 4 (2004) pp. 1–7. Available at http://www.tx.ncsu.edu/jtatm/volume4issue1/articles/Hinestroza/hinestroza_full_93_04.pdf (accessed 15 February 2011).
[10] C.T. Huang, C.L. Shen, C.F. Tang and S.H. Chang, Sens. Actuators, A 141 (2008) pp. 396–403.
[11] S. Mondal, Appl. Therm. Eng. 28(11–12) (2008) pp. 1536–1550.
[12] M.G. Lines, J. Alloy Compd. 449 (2008) pp. 242–245.
[13] B.S. Flavel, J. Yu, J.G. Shapter and J.S. Quinton, Carbon 45 (2007) pp. 2551–2558.
[14] J.M. Gutiérrez, C. González, A. Maestro, I. Solè, C.M. Pey and J. Nolla, Curr. Opin. Colloid Interf. Sci. 13 (2008) pp. 245–251.
[15] V.K. Srivastava, G. Kini and D. Rout, J. Colloid Interf. Sci. 304 (2006) pp. 214–221.
[16] S.B. Todorova, C.J.S.M. Silva, N.P. Simeonov and A.C. Paulo, Enzyme Microb. Technol. 40 (2007) pp. 1646–1650.
[17] C. Solans, P. Izquierdo, J. Nolla, N. Azemar and M.J.G. Celma, Curr. Opin. Colloid Interf. Sci. 10 (2005) pp. 102–110.
[18] M. Joshi, *The impact of nanotechnology in polyesters and polyamides*, in *Polyesters and Polyamides*, B.L. Deopura, R. Alagirusamy, M. Joshi and B. Gupta, eds., Woodhead, Cambridge, UK, 2008, pp. 354–415.
[19] M.L. Gulurajani, Indian J. Fiber Text. Res. 31 (2006) pp. 181–201.
[20] D.S. Soane and D.A. Offord, *WO patent nos. 9949124, 9949125*. Avantgarb LLC, USA, 30 September 1999.
[21] D.S. Soane, D.A. Offord, M.R. Linford and W. Ware Jr., *WO patent no. 0118303*. Nanotex LLC, USA, 15 March 2001.
[22] M.A. Kader and C. Nah, Polymer 45 (2004) pp. 2237–2247.
[23] X. Zhang, M. Järn, J. Peltonen, V. Pore, T. Vuorinen, E. Levänen and T. Mäntylä, J. Eur. Ceram. Soc. 28 (2008) pp. 2177–2181.
[24] K.C. Chang, Y.K. Chen and H. Chen, Surf. Coat. Tech. 201 (2007) pp. 9579–9586.
[25] M. Yu, G. Gu, W.D. Meng and F.L. Qing, Appl. Surface Sci. 253 (2007) pp. 3669–3673.
[26] W.A. Daoud and J.H. Xin, Sol-Gel Technol. 29 (2004) pp. 25–29.
[27] W.A. Daoud and J.H. Xin, J. Am. Ceram. Soc. 87 (2004) pp. 953–955.
[28] W.A. Daoud and J.H. Xin, Chem. Commun. 16 (2005) pp. 2110–2112.

[29] A. Bozzi, T. Yuranova, I. Guasaquillo, D. Laub and J. Kiwi, J. Photochem. Photobiol. A: Chem. 174 (2005) pp. 156–164.
[30] Y. Dong, Z. Bai, R. Liu and T. Zhu, Atmos. Environ. 41 (2007) pp. 3182–3192.
[31] T. Yuranova, R. Mosteo, J. Bandara, D. Laub and J. Kiwi, J. Mol. Catal. A: Chem. 244 (2006) pp. 160–167.
[32] http://www.nanotex.com/technologies/coolest_comfort.html (accessed 15 February 2011).
[33] S.H. Jeong, S.Y. Yeo and S.C. Yi, J. Mat. Sci. 40 (2005), pp. 5407–5411.
[34] H.J. Lee, S.Y. Yeo and S.H. Jeong, J. Mat. Sci. 38 (2003) pp. 2199–2204.
[35] H.Y. Ki, J.H. Kim, S.C. Kwon and S.H. Jeong, J Mat. Sci. 42 (2007) pp. 8020–8024.
[36] I. Sondi and B.S. Sondi, J. Colloid Interf. Sci. 275 (2004) pp. 177–182.
[37] X. Chen and H.J. Schluesener, Toxicol. Lett. 176 (2008) pp. 1–12.
[38] S.A. Blaser, M. Scheringer, M. MacLeod and K. Hungerbühler, Sci. Total Environ. 390 (2008) pp. 396–409.
[39] W. Dierickx and P.V.D. Berghe, Geotext. Geomembr. 22 (2004) pp. 255–272.
[40] A. Yadav, V. Prasad, A.A. Kathe, S. Raj, D. Yadav, C. Sundaramoorthy and N. Vigneshwaran, Bull. Mat. Sci. 29 (2006) pp. 641–645.
[41] Z. Zhou, L. Chu, W. Tang and L. Gu, J. Electrostat. 57 (2003) pp. 347–354.
[42] Y. Wu, Y.B. Chi, J.X. Nie, A.P. Yu, X.H. Chen and H.C. Ghu, J. Func. Polym. 15 (2002) pp. 42–47.
[43] P. Xu, W. Wang and S.L. Chen, Melliand Int. II (2005) pp. 56–59.
[44] J.E. McIntyre and P.N. Daniels (eds.), *Textile Terms and Definitions*, 10th ed., Textile Institute, Manchester, UK, 1995, pp. 67.
[45] I. Holme, Int. J. Adhes. Adhes. 19 (1999) pp. 455–463.
[46] G.B. Bantchev, Z. Lu, S. Eadula, M. Agrawal, G. Grozdits and Y. Lvov, *Layer-by-layer modification of lignocellulose fibers for better paper, Proceedings of the 62nd Southwest Regional Meeting of American Chemical Society*, Houston, TX, 2006.
[47] S.T. Dubas, P. Kumlangdudsana and P. Potiyaraj, Colloids Surf., A 289 (2006) pp. 105–109.
[48] K. Hyde, M. Rusa and J. Hinestroza, Nanotechnology 16 (2005) pp. 5422–5428.
[49] D.R. Baer, P.E. Burrows and A.A.E. Azab, Prog. Org. Coat. 47 (2003) pp. 342–356.
[50] H. Höcker, Pure Appl. Chem. 74 (2002) pp. 423–427.
[51] A. Sparavigna, *Plasma Treatment Advantages for Textiles*. Available at http://arxiv.org/ftp/arxiv/papers/0801/0801.3727.pdf (accessed 15 February 2011).
[52] F. Leroux, C. Campagne, A. Perwuelz and L. Gengembre, Appl. Surface Sci. 254 (2008) pp. 3902–3908.
[53] A. Bozzi, T. Yuranova and J. Kiwi, J. Photochem. Photobiol. A: Chem. 172 (2005) pp. 27–34.
[54] J.M.P. Hernandez, J. Manriquez, Y.M. Vong, F.J. Rodriguez, T.W. Chapman, M.I. Maldonado and L.A. Godinez, J. Hazard. Mat. 147 (2007) pp. 588–593.
[55] R.A. Caruso and M. Antomitte, Chem. Mater. 13 (2001) pp. 3272–3282.
[56] B. Xu and Z. Kai, Appl. Surface Sci. 254 (2008) pp. 5899–5904.
[57] B.G. Prevo, D.M. Kuncicky and O.D. Velev, Colloids Surf., A 311 (2007) pp. 2–10.
[58] K.T. Meilert, D. Laub and J. Kiwi, J. Mol. Catal. A: Chem. 237 (2005) pp. 101–108.
[59] S. Wazed Ali, S. Rajendran and M. Joshi, Polym. Polym. Compos. 18 (2010) pp. 237–249.
[60] L. Zhai, F.C. Cebeci, R.E. Cohen and M.F. Rubner, Nanoletters 4 (2004) pp. 1349–1353.
[61] M. Joshi, R. Khanna, R. Shekhar and K. Jha, J. Appl. Polym. Sci. 119 (2011) pp. 2793–2499.
[62] E.G. Han, E.A. Kim and K.W. Oh, Synth. Met. 123 (2001) pp. 469–476.
[63] D. Kincal, A. Kumar, A.D. Child and J.R. Reynolds, Synth. Met. 92 (1998) pp. 53–56.
[64] M. Jaiswal and R. Menon, Polym. Int. 55 (2006) pp. 1371–1384.
[65] T. Lin, L. Wang, X. Wang and A. Kaynak, Thin Solid Films 479 (2005) pp. 77–82.
[66] A. Varesano, L.D. Acqua and C. Tonin, Polym. Degrad. Stab. 89 (2005) pp. 125–132.
[67] J. Wu, D. Zhou, C.O. Too and G.G. Wallace, Synth. Met. 155 (2005) pp. 698–701.
[68] L.D. Acqua, C. Tonin, A. Varesano, M. Canetti, W. Porzio and M. Catellani, Synth. Met. 156 (2006) pp. 379–386.
[69] A. Malinauskas, Polymer 42 (2001) pp. 3957–3972.
[70] H.H. Kuhn, A.D. Child and W.C. Kimbrell, Synth. Met. 71 (1995) pp. 2139–2142.
[71] E. Hu, A. Kaynak and Y. Li, Synth. Met. 150 (2005) pp. 139–143.
[72] W. Fung, *Coated and Laminated Textiles*, Woodhead, Cambridge, UK, 2002.
[73] A.K. Sen, *Coated Textiles: Principles and Applications*, Woodhead, Cambridge, UK, 2001.

[74] M.E. Hall, *Coating of technical textiles*, in *Handbook of Technical Textiles*, A.R. Horrocks and S.C. Anand, eds., Woodhead, Cambridge, UK, 2000, pp. 173–186.

[75] C.T. Nguyen, T.V. Khanh and J. Lara, Theor. Appl. Fract. Mech. 42 (2004) pp. 25–33.

[76] H.G. Schmelzer, Mat. Design 9 (1988) pp. 276–286.

[77] D.K. Chattopadhyay and K.V.S.N. Raju, Prog. Polym. Sci. 32 (2007) pp. 352–418.

[78] X. Chen, Y. Hu, C. Jiao and L. Song, Polym. Degrad. Stab. 92 (2007) pp. 1141–1150.

[79] X. Hao, J. Zhang and Y. Guo, Eur. Polym. J. 40 (2004) pp. 673–678.

[80] G. Budden, J. Indust. Text. 34 (2004) pp. 117–125.

[81] M. Rochery, I. Vroman and C. Campagne, J. Indust. Text. 35 (2006) pp. 227–238.

[82] Y. Okamoto, Y. Hasegawa and F. Yoshino, Prog. Org. Coat. 29 (1996) pp. 175–182.

[83] B. Boutevin and Y. Pietrasanta, Prog. Org. Coat. 13 (1985) pp. 297–331.

[84] A. Jain, J.S. Gutmann, C.B.W. Garcia, Y. Zhang, M.W. Tate, S.M. Gruner and U. Wiesner, Macromolecules 35 (2002) pp. 4862–4865.

[85] H.H. Murray, Appl. Clay Sci. 17 (2000) pp. 207–221.

[86] Y.W.C. Yang, Y.K. Lee, Y.T. Chen and J.C. Wu, Polymer 48 (2007) pp. 2969–2979.

[87] M. Zbik and R.G. Horn, Colloids Surf., A 222 (2003) pp. 323–328.

[88] F. Bergaya and G. Lagaly, Appl. Clay Sci. 19 (2001) pp. 1–3.

[89] M. Alexandre and P. Dubois, Mater. Sci. Eng. 28 (2000) pp. 1–63.

[90] S.S. Ray and M. Okamoto, Prog. Polym. Sci. 28 (2003) pp. 1539–1641.

[91] D. Burgentzlé, J. Duchet, J.F. Gérard, A. Jupin and B. Fillon, J. Colloid Interf. Sci. 278 (2004) pp. 26–39.

[92] P.C. LeBaron, Z. Wang and T.J. Pinnavaia, Appl. Clay Sci. 15 (1999) pp. 11–29.

[93] F.A. Bergaya, Micropor. Mesopor. Mater. 107 (2008) pp. 141–148.

[94] Q. Liu, J.R.D. Wijn and C.A.V. Blitterswijk, Biomaterials 18 (1997) pp. 1263–1270.

[95] K.C. Chang, S.T. Chen, H.F. Lin, C.Y. Lin, H.H. Huang, J.M. Yeh and Y.H. Yu, Eur. Polym. J. 44 (2008) pp. 13–23.

[96] D.A. Brune and J. Bicerano, Polymer 43 (2002) pp. 369–387.

[97] J.M. Gloaguen and J.M. Lefebvre, Polymer 42 (2001) pp. 5841–5847.

[98] S. Subramani, S.W. Choi, J.Y. Lee and J.H. Kim, Polymer 48 (2007) pp. 4691–4703.

[99] S. Subramani, J.Y. Lee, J.H. Kim and I.W. Cheong, Compos. Sci. Tech. 67 (2007) pp. 1561–1573.

[100] E. Picard, A. Vermogen, J.F. Gerard and E. Espuche, J. Membr. Sci. 292 (2007) pp. 133–144.

[101] S. Takahashi, H.A. Goldberg, C.A. Feeney, D.P. Karim, M. Farrell, K. O'Leary and D.R. Paul, Polymer 47 (2006) pp. 3083–3093.

[102] M. Joshi, K. Banerjee, R Prasanth and V. Thakare, Indian J. Fibre Text. Res. 31 (2006) pp. 202–214.

[103] C.D. Muzny, B.D. Butler, H.J.M. Hanley, F. Tsvetkov and D.G. Peiffer, Mater. Lett. 28 (1996) pp. 379–384.

[104] Z.Y. Wang, E.H. Han and W. Ke, J. Appl. Polym. Sci. 103 (2007) pp. 1681–1689.

[105] A.R. Horrocks, B.K. Kandola, P.J. Davies, S. Zhang and S.A. Padbury, Polym. Degrad. Stab. 88 (2005) pp. 3–12.

[106] Z.H. Chang, F. Guo, J.F. Chen, J.H. Yu and G.Q. Wang, Polym. Degrad. Stab. 92 (2007) pp. 1204–1212.

[107] J.H. Chang and Y.U. An, J. Polym. Sci.: Part B: Polym. Phy. 40 (2002) pp. 670–677.

[108] K.J. Yao, M. Song, D.J. Hourston and D.J. Luo, Polymer 43 (2002) pp. 1017–1020.

[109] Y.I. Tien and K.H. Wei, Polymer 42 (2001) pp. 3213–3221.

[110] S. Kumari, A.K. Mishra, A.V.R. Krishna and K.V.S.N. Raju, Prog. Org. Coat. 60 (2007) pp. 54–62.

[111] A. Okada and A. Usuki, Mater. Sci. Eng., C 3 (1995) pp. 109–115.

[112] C.E. Corcione, P. Prinari, D. Cannolettaa, G. Mensitieri and A. Maffezzoli, Int. J. Adhes. Adhes. 28 (2008) pp. 91–100.

[113] J. Xiong, Z. Zheng, H. Jiang, S. Ye and X. Wang, Composites Part A 38 (2007) pp. 132–137.

[114] T.K. Chen, Y.I. Tien and K.H. Wei, Polymer 41 (2000) pp. 1345–1353.

[115] J. Zheng, R. Ozisik and R.W. Siegel, Polymer 47 (2006) pp. 7786–7794.

[116] F. Chavarria and D.R. Paul, Polymer 47 (2006) pp. 7760–7773.

[117] A. Pattanayak and S.C. Jana, Polymer 46 (2005) pp. 3394–3406.

[118] A. Pattanayak and S.C. Jana, Polymer 46 (2005) pp. 3275–3288.

[119] B.K. Kim, J.W. Seo and H.M. Jeong, Eur. Polym. J. 39 (2003) pp. 85–91.

[120] F. Cao and S.C. Jana, Polymer 48 (2007) pp. 3790–3800.
[121] Y.H. Yu, J.M. Yeh, S.J. Liou and Y.P. Chang, Acta Mater. 52 (2004) pp. 475–486.
[122] M.R. Bagherzadeh and F. Mahdavi, Prog. Org. Coat. 60 (2007) pp. 117–120.
[123] G.Y. Bae, B.G. Min, Y.G. Jeong, S.C. Lee, J.H. Jang and G.H. Koo, J. Colloid Interf. Sci. 337 (2009) pp. 170–175.
[124] C.H. Xue, S.T. Jia, J. Zhang and L.Q. Tian, Thin Solid Films 517 (2009) pp. 4593–4598.
[125] Y.W.H. Wong, C.W.M. Yuen, M.Y.S. Leung, S.K.A. Ku and H.L.I. Lam, AUTEX Res. J. 6 (2006) pp. 1–8. Available at http://www.freewebs.com/jayaram-co/doc/Selected_Appz_of_Nanotechnology_in_Textiles.pdf (accessed 15 February 2011).
[126] M. Garcı, M. de Rooij, L. Winnubst, W.E. van Zyl and H. Verweij, J. Appl. Polym. Sci. 92 (2004) pp. 1855–1862.
[127] H.J. Song, Z.Z. Zhang and X.H. Men, Composites Part A 39 (2008) pp. 188–194.
[128] T. Kashiwagi, A.B. Morgan, J.M. Antonucci, M.R. VanLandingham, R.H. Harris Jr., W.H. Awad and J.R. Shields, J. Appl. Polym. Sci. 89 (2003) pp. 2072–2078.
[129] N.G. Sahoo, Y.C. Jung, H.J. Yoo and J.W. Cho, Compos. Sci. Tech. 67 (2007) pp. 1920–1929.
[130] Z.Z. Zhang, H.J. Song, X.H. Men and Z.Z. Luo, Wear 264 (2008) pp. 599–604.
[131] P.C. Ma, J.K. Kim and B.Z. Tang, Carbon 44 (2006) pp. 3232–3238.
[132] H.J. Song, Z.Z. Zhang and X.H. Men, Eur. Polym. J. 43 (2007) pp. 4092–4102.
[133] T.L. Makarova, Diamond Relat. Mater. 16 (2007) pp. 1841–1846.
[134] A.I. Romanenko, O.B. Anikeeva, V.L. Kuznetsov, A.N. Obrastsov, A.P. Volkov and A.V. Garshev, Solid State Commun. 137 (2006) pp. 625–629.
[135] Y.C. Huang, Physica Status Solidi C 4 (2007) pp. 540–543.
[136] J. Walter, S. Wakita, W. Boonchuduang and S. Hara, J. Phys. Chem. B 106 (2002) pp. 8547–8554.
[137] A. Bhattacharyya and M. Joshi, J. Nanostruc. Polym. Nanocompos. 6 (2010) pp. 73–78.
[138] S. Duquesne, M.L. Bras, S. Bourbigot, R. Delobel, G. Camino, B. Eling, C. Lindsay and T. Roels, Polym. Degrad. Stab. 74 (2001) pp. 493–499.
[139] S. Choi and B.V. Sankar, Composites Part B 39 (2008) pp. 782–791.
[140] T. Enoki, J. Phys. Chem. Solids 65 (2004), pp. 103–108.
[141] S. Giraud, S. Bourbigot, M. Rochery, I. Vroman, L. Tighzert, R. Delobel and F. Poutch, Polym. Degrad. Stab. 88 (2005) pp. 106–113.
[142] P. Katangur, P.K. Patra and S.B. Warner, Polym. Degrad. Stab. 91 (2006) pp. 2437–2442.
[143] Z. Wang, E. Han and W. Ke, Surf. Coat. Tech. 200 (2006) pp. 5706–5716.
[144] C. Friedrich, G. Berg, E. Broszeit and C. Berger, Surf. Coat. Tech. 98 (1998) pp. 816–822.
[145] Y. Wang, S. Lim, J.L. Luo and Z.H. Xu, Wear 260 (2006) pp. 976–983.
[146] L. Chang, Z. Zhang, H. Zhang and A.K. Schlar, Compos. Sci. Tech. 66 (2006) pp. 3188–3198.
[147] S.H. Hsu and C.W. Chou, Polym. Degrad. Stab. 85 (2004) pp. 675–680.
[148] J. Chen, M. Liu, L. Zhang, J. Zhang and L. Jin, Water Res. 37 (2003) pp. 3815–3820.
[149] Q. Fan, S.C. Ugbolue, A.R. Wilson, Y.S. Dar and Y. Yang, *Dyeable polypropylene via nanotechnology*, NTC project C01-MD20. Available at http://www.ntcresearch.org/pdf-rpts/Bref0602/C01-MD20-02.pdf (accessed 15 February 2011)
[150] L. Razafimahefa, S. Chlebicki, I. Vroman and E. Devaux, Dyes Pigm. 66 (2005) pp. 55–60.
[151] C. Wang, K. Fang and W. Ji, Fibers Polym. 8 (2007), pp. 225–229.
[152] J. Yuan, S. Zhou, G. Gu and L. Wu, J. Sol-Gel Sci. Tech. 36 (2005) pp. 265–274.
[153] D. Hegemann, M.M. Hossain and D.J. Balazs, Prog. Org. Coat. 58 (2007) pp. 237–240.
[154] J. Jang and Y. Jeong, Dyes Pigm. 69 (2006) pp. 137–143.
[155] P.M. Ajayan, L.S. Schadler and P.V. Braun (eds.), *Nanocomposite Science and Technology*, Willey VCH, Germany, 2003.
[156] D.W. Schaefer and J.E. Mark (eds.), *Polymer-Based Molecular Nanocomposites*, Materials Research Society, Pittsburgh, PA, 1990.
[157] A. Okada and A. Usuki, Macromol. Mat. Eng. 291 (2006) pp. 1449–1476.
[158] S. Brauer, *Polymer Nanocomposites: Nanoparticles, Nanoclays and Nanotubes*, report ID Nano21B, Business Communications Company (BCC), Wellesley, MA, 2004, pp. 200.
[159] R.A. Vaia, AMPTIAC Q. 6 (2002), pp. 17–24.
[160] L.A. Utracki, *Clay-Containing Polymeric Nanocomposites*, RAPRA, Shawbury, Shropshine, UK, 2004.
[161] L.A. Utracki, Indian J. Fibre Text. Res. 31 (2006) pp. 15–28.

[162] Y.K. Kim, A.F. Lewis, P.K. Patra, S.B. Warner, S.K. Mhetre, M.A. Shah and D. Nam, Proc. Mater. Res. Soc. Symp. 740 (2002) pp. 441–446.

[163] S.J. Kim, M.U. Mun, J.H. Chang, Polymer (Korea) 29 (2005), pp. 190–197.

[164] W. Xiao, H. Yu, K. Han and M. Yu, J. Appl. Polym. Sci. 96 (2005) pp. 2247–2252.

[165] I.H. Kwon, Y.H. Bang and D.J. Lee, *Korean patent 2005 5090851*. Hyosung Corporation, S. Korea, 2005.

[166] Y. Yang and H. Gu, *Chinese patent CN1760443*. Shanghai Jiao Tong University, People's Republic of China, 2006.

[167] J.H. Chang, M.K. Mun and J.C. Kim, J. Appl. Polym. Sci. 102 (2006) pp. 4535–4545.

[168] K. Yang and R. Ozisik, Polymer 47 (2006) pp. 2849–2855.

[169] D.H. Seo, Y.M. Lee and J.D. Nam, *Korean patent 2006 060454*. Garamtech Co. Ltd., S Korea, 2006.

[170] K.H. Yoon, M.B. Polk, B.G. Min and D.A. Schiraldi, Polym. Int. 53 (2004) pp. 2072–2078.

[171] G. Chen, D. Shen, M. Feng and M. Yang, Macromol. Rapid Comm. 25 (2004) pp. 1121–1124.

[172] M.G. McCord, S.R. Matthews and S.M. Hudson, J. Adv. Mat. 36 (2004), pp. 44–56.

[173] J.S. Dugan, *US Pat. 2002 110686 (A1)*. Fiber Innovation Technology, Inc. USA, 2002.

[174] S. Bourbigot, E. Devaux and X. Flambard, Polym. Degrad. Stab. 75 (2002), pp. 397–402.

[175] W. Kowbel, K. Patel and J.C. Withers, *Nanocomposites for school bus upholstery*, in *Proceedings of the International SAMPE Symposium Exhibition*, Society for the Advancement of Material and Process Engineering, A Materials and Processes Odyssey, Book 2 46, Long Beach, CA, May 2001, pp. 2577–2581.

[176] E. Giza, K. Hiroshi, T. Kikutani and N.O. Kui, J. Polym. Eng. 20 (2000) pp. 403–425.

[177] E. Giza, K. Hiroshi, T. Kikutani and N.O. Kui, J. Macromol. Sci. Phys. B39 (2000) pp. 545–559.

[178] C. Ibanes, M.D. Boissieu, C. David and R. Seguela, Polymer 47 (2006) pp. 5071–5079.

[179] Y.H. Lee and B.J. Kim, *Korean patent KR2006078370 (AN: 1259608)*. Hyosung Corporation, S Korea, 2006.

[180] J. Ma and B. Liang, *Chinese patent CN1587460 A 20050302 (AN: 131581)*. Donghua University, People's Republic, China, 2006.

[181] J.H. Chang, M.K. Mun and I.C. Lee, J. Appl. Polym. Sci. 98 (2005) pp. 2009–2016.

[182] M.K. Mun, J.C. Kim and J.H. Chang, Polym. Bull. (Heidelberg, Germany) 57 (2006), pp. 797–804.

[183] G.H. Guan, C.C. Li and D. Zhang, J. Appl. Polym. Sci. 95 (2005), pp. 1443–1447.

[184] J.H. Chang and S.J. Kim, Polym. Bull. (Heidelberg, Germany) 52 (2004) pp. 289–296.

[185] C.D. Delhom, *Development and Thermal Characterization of Cellose/Clay Nanocomposites*. Available at http://etd.lsu.edu/docs/available/etd-04032009-094316/unrestricted/Delhom-ETD-Final.pdf (accessed 15 February 2011).

[186] G.V.R. Reddy, B.L. Deopura and M. Joshi, *Dry-Jet-Wet Spun Polyurethane/Clay Nanocomposite Filaments*. Available at http://www.thefibersociety.org/Assets/Past_Meetings/BooksOfAbstracts/2007_Spr_BookAbstracts.pdf (accessed 15 February 2011).

[187] M. Joshi and V. Viswanathan, J. Appl. Polym. Sci. 102 (2006) pp. 2164–2174.

[188] M. Joshi, M. Shaw and B.S. Butola, Fibers Polym. 5 (2004) pp. 59–67.

[189] M. Moniruzzaman and K.I. Winey, Macromolecules 39 (2006) pp. 5194–5204.

[190] P.M. Ajayan, L.S. Schadhler, C. Giannaris and A. Rubio, Adv. Mater. 12 (2000) pp. 401–411.

[191] P.M. Ajayan, O. Stephan, C. Colliex and D. Trauth, Science 265 (1994) pp. 1212–1214.

[192] G.C. Han and S. Kumar, Indian J. Fibre Text. Res. 31 (2006) pp. 29–40.

[193] C.A. Dyke and J.M. Tour, Chem. Eur. J. 10 (2004) pp. 812–817.

[194] A. Star, Y. Lui, K. Grant, L. Ridvan, J.F. Stoddart, D.W. Steuerman, M.R.M. Diehl, A. Boukai and J.R. Heith, Macromolecules 36 (2003) pp. 553–560.

[195] A.B. Dalton, S. Collins, E. Munoz, J.M. Razal, V.H. Ebron, J.P. Ferraris, J.N. Coleman, B.G. Kim and R.H. Baughman, Nature 423 (2003) pp. 703–703.

[196] Y. Kim, A.F. Lewis, S.B. Warner, P.K. Patra and P.D. Calvert, *Nanocomposite Fibers*. Available at http://www.ntcresearch.org/pdf-rpts/AnRp03/M00-MD08-A3.pdf (accessed 15 February 2011).

[197] H. Meng, G.X. Sui, P.F. Fang and R. Yang, Polymer 49 (2008) pp. 610–620.

[198] M. Jose, J. Tyner, E. Nyairo, D. Dean, *Synthesis and processing of aligned carbon nanotube-based fibers*, *Proceedings of the 49th International SAMPE Symposium and Exhibition Conference*, Long Beach, CA, May 2004.

[199] J.P. Thomas, A. Rohatgi, J.N. Baucom and W.R. Pogue, *Preliminary investigations of a hierarchically structured nanocomposites, Proceedings of the American Chemistry Society for Composites*, 20th Technical Conference, Birmingham, AL, 2005 (35/1-35/10.A 2005:1280403).

[200] W. Chen, X. Tao and Y. Liu, Comp. Sci. Tech. 66 (2006) pp. 3029–3034.

[201] M. Mahfuz, A. Adanan, V. Rangari and S. Julani, Int. J. Nanosci. 4 (2005) pp. 55–72.

[202] M. Zhu, Q. Xiang, H. He, Y. Zhang, P. Yanmo, P. Poetschke and H.J. Adler, Macromol. Symp. 210 (2004) pp. 251–261 (AN 2004:41400).

[203] Y. Yang and H. Gu, J. Appl. Polym. Sci. 105 (2007) pp. 2363–2369.

[204] Y. Yang and H. Gu, J. Appl. Polym. Sci. 102 (2006) pp. 3691–3697.

[205] P. Cui, J. Jin, X. Tian, H. Zhang, Y. Li, W. Lui, K. Zheng, J. Zheng, Z. Meng and Q. Wu, *Chinese patent CN 1004-1160 20040702, AN2006: 136051*, 2006.

[206] K. Han and M. Yu, J. Appl. Polym. Sci. 100 (2006) pp. 1588–1593.

[207] X. Xu, Y. Guo, W. Xu, A. Zhang, G. Ye and P. Liu, *Chinese patent CN 2001-108688, AN 2004:519148*. Sichuan University, People's Republic of China, 2004.

[208] M. Joshi and B.S. Butola, J. Appl. Polym. Sci. 105 (2007) pp. 978–985.

[209] G. Li, L. Wang, H. Niand and C.U. Pittman, Jr., J. Inorg. Organomat. Polym. 11 (2001) pp. 123–154.

[210] M. Joshi and B.S. Butola, J. Macromol. Sci.: Part-C Polym. Rev. 44 (2004) pp. 389–410.

[211] K.H. Yoon, M.B. Polk, J.H. Park, B.G. Min and D.A. Schiraldi, Polym. Int. 54 (2005) pp. 47–53.

[212] J. Zeng, S. Kumar, S. Iyar, D.A. Schiraldi and R.I. Gonzalez, High Perform. Polym. 17 (2005) pp. 403–424.

[213] A. Bhattacharyya and M. Joshi, *Electroactive polyurethane nanocomposite fibers for intelligent textile applications, Proceedings of the 25th Annual Meeting of Polymer Processing Society* (PPS-25 Goa), March 2009, GS-VII: PP6.

[214] K. Graham, G.H. Schreuder and M. Gogins, *Incorporation of electrospun nanofibers into functional structures, Proceedings of the Non-Woven Technical Conference*, Baltimore, September 2003, pp. 15–18.

[215] Y.M. Shin, M.M. Hohman, M.P. Brenner and G.C. Rutledge, Appl. Phys. Lett. 78 (2001) pp. 1149–1151.

[216] Y.M. Shin, M.M. Hohman, M.P. Brenner and G.C. Rutledge, Polymer 42 (2001) pp. 9955–9967.

[217] A. Formhals, *US patent no. 364780*, 1932.

[218] D.H. Renekar and I. Chun, Nanotechnology 7 (1996) pp. 216–233.

[219] T. Subbiah, G.S. Bhatt, R.W. Tock, S. Paraneswaran and S. S Ramkumar, J. Appl. Polym. Sci. 96 (2005) pp. 557–569.

[220] Z.M. Huang, Y.Z. Zhang, M. Kotaki and S. Ramakrishna, Comp. Sci. Tech. 63 (2003) pp. 2223–2253.

[221] S. Megelski, J.S. Stephans, D.B. Chase and J.F. Rabolt, Macromolecules 35 (2002) pp. 8456–8466.

[222] M. Deitzel, J. Kleinmayer, D. Harris and N.C. Beck Tan, Polymer 42 (2001) pp. 261–272.

[223] A. Buer, S.C. Ugbolue and S.B. Warner, Text. Res. J. 71 (2001) pp. 323–328.

[224] G. Srinivasan and D.H. Renekar, Polym. Int. 36 (1995) pp. 195–201.

[225] H. Fong, I. Chun and D.H. Reneker, Polymer 40 (1999) pp. 4585–4592.

[226] L. Larrondo and R.St.J. Manley, J. Polym. Sci.: Polym. Phys. 19 (1981) pp. 909–920.

[227] L. Larrondo and R.St.J. Manley, J. Polym. Sci.: Polym. Phys. 19 (1981) pp. 921–932.

[228] A.L. Yarin, S. Koombhongse and D.H. Reneker, J. Appl. Phys. 90 (2001) pp. 4836–4846.

[229] L. Larrondo and R.St.J. Manley, J. Polym. Sci.: Polym. Phys. 19 (1981) pp. 933–940.

[230] P.K. Baumgarten, J. Colloid Interf. Sci. 36 (1971) pp. 71–79.

[231] D.H. Reneker, A.L. Yarin, H. Fong and S. Koombhongse, J. Appl. Phys. 87 (2000) pp. 4531–4547.

[232] A.L. Yarin, S. Koombhongse and D.H. Reneker, J. Appl. Phys. 89 (2001) pp. 3018–3026.

[233] J. Kamioka and H.G. Craighead, Appl. Phys. Lett. 83 (2003) pp. 371–373.

[234] B. Sundaray, V. Subramanian, T.S. Nataraja, R.Z. Xiang, C. Chiacheng and W.S. Fam, Appl. Phys. Lett. 84 (2004) pp. 1212–1224.

[235] D. Li, Y. Wang and Y. Xia, Nanoletters 3 (2003) pp. 1167–1171.

[236] M.M. Demir, I. Yilgor, E. Yilgor and B. Erman, Polymer 43 (2002) pp. 3303–3309.

[237] J. Doshi and D.H. Reneker, J. Electrostat. 35 (1995) pp. 151–160.

[238] C.S. Ki, D.H. Back, K.D. Gang, K.H. Lee, I.C. Um and Y.H. Park, Polymer 46 (2005) pp. 5094–5102.

[239] C.J. Buchko, L.C. Chen, Y. Shen and D.C. Martin, Polymer 40 (1999) pp. 7397–7407.

[240] K.H. Lee, H.Y. Kim, M.S. Khil, Y.M. Ra and D.R. Lee, Polymer 44 (2003) pp. 1287–1294.

[241] K.J. Pawlowski, H.L. Belvin, D.L. Raney, J. Su, J.S. Harrison and E.J. Siochi, Polymer 44 (2003) pp. 1309–1314.

[242] S. Koombhongse, W. Liu and D.H. Reneker, J. Polym. Sci.: Polym. Phys. 39 (2001) pp. 2598–2606.

[243] J. Lyons, C. Li and F. Ko, Polymer 45 (2004), pp. 7597–7603.

[244] S. Warner, A. Fowler, S. Ugbolue and M. Jaffe, *Cost Effective Nanofiber Formation – Melt Spinning*, NTC project F05-MD01, 2007. Available at http://www.ntcresearch.org/pdf-rpts/AnRp05/F05-MD01-A5.pdf. (accessed 15 February 2011).

[245] E.R. Kenawy, J.M. Layman, J.R. Watkins, G.L. Bowlin, J.A. Matthews, D.G. Simpson and G.E. Wnek, Biomaterials 24 (2003) pp. 907–913.

[246] C. Nah, S.H. Han, M.H. Lee, J.S. Kim and D.S. Lee, Polym. Int. 52 (2003) pp. 429–432.

[247] K. Ohgo, C. Zhao, M. Kobayashi and T. Asakura, Polymer 44 (2003), 841–846.

[248] C. Shao, H.Y. Kim, J. Gong, B. Ding, D.R. Lee and S.J. Park, Mater. Lett. 57 (2003) pp. 1579–1584.

[249] C. Ulrich, Hum. Ecol. 34 (2006) pp. 2–5. Available at http://nanotextiles.human.cornell.edu/Nanotechnology%20in%20textiles%20-%20protective%20fibers.pdf (accessed 15 February 2011).

[250] M.M. Hussain and S.S. Ramkumar, Indian J. Fibre Text. Res. 31 (2006) pp. 41–51.

[251] H. Fong, W. Liu, C.S. Wang and R.A. Vaia, Polymer 43 (2002) pp. 775–780.

[252] N. Ristolainen, P. Heikkila, A. Harlin and J. Seppala, Macromol. Mater. Eng. 291 (2006) pp. 114–122.

[253] A. Baji, Y.W. Mai, S.C. Wong, M. Abtahi and X. Du, Comp. Sci. Tech. 70 (2010) pp. 1401–1409.

[254] Y.L. Zoo and H. Zhou, *US patent 7083854 B1 20060801*, 2006.

[255] M. Yang, G. Hu and S. Zhang, *Chinese patent CN 1001–1772, 20050503, AN 2006:1265080*, 2006.

[256] T.H. Graefe and K.M. Graham, *Nanofiber webs from electrospinning, Proceedings of the 5th International Conference on Nonwovens in Filtration*, Stuttgart, Germany, March 2003. Available at http://www.donaldson.com/en/filtermedia/support/datalibrary/052026.pdf (accessed 15 February 2011).

[257] J.M. Deitzel, W. Kosik, S.H. McKnight, N.C. Beck Tan, J.M. DeSimone and S. Crette, Polymer 43 (2002) pp. 1025–1029.

[258] S.A. Theron, A.L. Yarin, E. Zussman and E. Kroll, Polymer 46 (2005) pp. 2889–2899.

[259] D. Lukas, A. Sarkar and P. Pokorny, J. Appl. Phys. 103 (2008) pp. 084309-1–084309-7.

[260] A.L. Yarin and E. Zussman, Polymer 45 (2004) pp. 2977–2980.

[261] L. Torobin and R. Findlow, *US patent, 6, 183, 670*, 6 February 2001.

[262] J.F. Hagewood, Int. Fiber J. December (2002) pp. 62–63.

[263] K.I. Jacob, M. Liu, M. Polk and S. Bechtel, NTC project: M04-GT11. Available at http://www.ntcreaserch.org (accessed 15 February 2011).

[264] Y. Dzenis, Science 304 (2004) pp. 1917–1919.

[265] K. Jayraman, M. Kotaki, Y.Z. Zhang, X.M. Mo and S. Ramakrishna, J. Nanosci. Nanotech. (2004) pp. 52–65.